Electrostatic Kinetic Energy Harvesting

Nanotechnologies for Energy Recovery Set

coordinated by
Pascal Maigné

Volume 3

Electrostatic Kinetic Energy Harvesting

Philippe Basset
Elena Blokhina
Dimitri Galayko

ISTE

WILEY

First published 2016 in Great Britain and the United States by ISTE Ltd and John Wiley & Sons, Inc.

ISTE Ltd
27-37 St George's Road
London SW19 4EU
UK

www.iste.co.uk

John Wiley & Sons, Inc.
111 River Street
Hoboken, NJ 07030
USA

www.wiley.com

Library of Congress Control Number: 2016930389

British Library Cataloguing-in-Publication Data
A CIP record for this book is available from the British Library
ISBN 978-1-84821-716-4

Contents

Preface

Miniaturization and efficiency are current trends in modern microelectronics. They will most likely continue to be for the next few years as we enter the age of the Internet of Things. Future technologies will rely on energy sources and, for this reason, energy harvesting is an extremely active, versatile and developing area that includes engineers and scientists from the field of electronics, microsystems and material science. While there are many different types of energy harvesting systems, we believe that electrostatic kinetic energy harvesters – devices that generate electricity from mechanical motion employing the capacitive (electrostatic) mechanism – are underrepresented in the literature. Although they are particularly compatible with microtechnology, they are also more complex compared to other kinetic energy harvesters.

This book is the summary of collaboration between three research groups from Université Paris-Sorbonne, University College of Dublin and Université Paris-Est on kinetic energy harvesters carried out between 2007 and 2015. Although this book is focused on energy harvesting employing the electrostatic transduction, we believe that this allowed us to write a complete and self-sufficient study. The book covers all aspects necessary to understand and design an efficient harvester, including linear and nonlinear resonators, electrostatic transaction principles, microfabrication processes and the design of conditioning electronics.

This book is primarily intended for Master's degree and PhD students who wish to discover the field of kinetic energy harvesting. It contains both chapters on fundamentals and chapters that present state-of-the-art results. We believe that some chapters will also be of interest to scientist and engineers involved in the design and development of kinetic energy harvesting.

Chapters 1, 2 and 7 discuss a capacitive energy harvester as a system, with additional chapters devoted to the operation in both the electrical and mechanical domains. Chapters 3 through 6 discuss mechanical aspects of harvesters, and Chapters 8 through 11 are devoted to electronic conditioning circuitry. We have made a choice to present the material at a relatively high level of abstraction, limiting the discussion to the aspects that have the most impact on the global operation of the harvester. While this choice does not allow an extended discussion on some practical aspects of implementation, we are certain that this study provides a deep enough understanding of the role and function of each component in an energy harvester.

We sincerely thank all the PhD students and postdoctoral researchers who have contributed to our collaboration and research under our supervision: Mahmood Ayyaz Paracha, Andrii Dudka, Raphaël Guillemet, Peter Harte, Francesco Cottone, Eoghan O'Riordan, Armine Karami, Mohammed Bedier and Yingxian Lu. In particular, we are grateful to Eoghan O'Riordan, Peter Harte, Armine Karami, Yingxian Lu and Mohamed Bedier for helping prepare this book.

Philippe BASSET
University-Paris-Est

Elena BLOKHINA
University College Dublin

Dimitri GALAYKO
Sorbonne Universities, UPMC University Paris 06
January 2016

Introduction: Background and Area of Application

Sensing and data recording is appearing as the new paradigm of the 21st Century: smart cars, smart homes, smart buildings and cities are the objects of very extensive and active research. They all need a large number of communicating sensors (preferably wireless) for installation and operation cost reduction, or reliability improvement. The modern car has a few hundred sensors, and it is expected that the automotive industry will require 22 billion sensors by 2020. Today, most sensors are still powered using wires. Making them autonomous would ease the sensor installation and would, generally, offer a lighter and more reliable system. These ideas about smart environments benefit from the wireless sensor network (WSN) or the Internet of Thing (IoT) concepts. In WSNs, all the sensors are provided with an embedded energy source and an antenna to wirelessly transmit data measurements. This communication system usually takes place in a star network where each sensor communicates with a master node. A better implementation of the network can be arranged if each sensor communicates with the closest node in order to progressively propagate the measured information to reach the base station (Figure I.1). For the IoT, the main idea is that any item of daily life is able to communicate data through such a network.

For both the WSN and the IoT, independent and miniature power sources are usually preferable over wire powering. In the majority of cases, a battery is used, which may last from several days to several years, depending on its size and the application. However, there are applications where a battery is not suitable for some reasons: a harsh environment degrades the battery too fast, an inaccessible location makes the cost of the battery replacement too high, or, for example, a chemical battery may be incompatible with ecological requirements. In these cases, an appropriate solution would be to take energy from the ambient environment of the sensor. This is the modern concept of "energy harvesting".

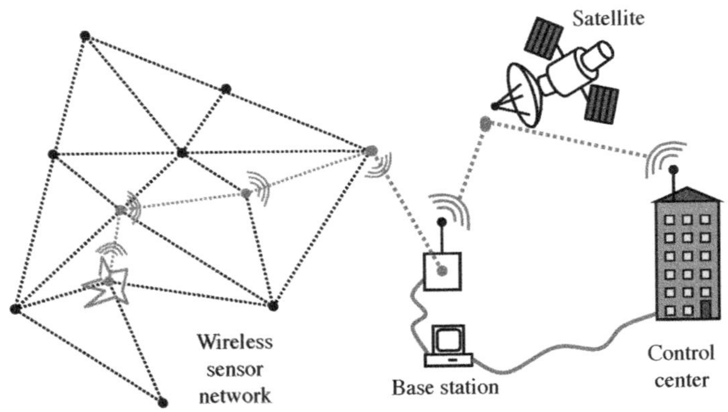

Figure I.1. *Principle of an ideal wireless sensor network (WSN) ecosystem: an event captured by an abandoned sensor node is transmitted through communications with other nodes nearby*

There are many ways to harvest ambient energy. The most mature and efficient technique is the use of photovoltaic cells converting ambient light into electricity. Another useful, but more complex technique, is the use of the Seebeck effect to obtain electrical power from a temperature gradient. However, in some cases where no light or temperature gradient are available in the environment, less conventional energy sources have to be envisaged. Kinetic energy, and more specifically vibrations, can provide great opportunities since these are present in the environment of many applications.

The idea of using kinetic energy to power a system is well known, such as with the dynamo invented in the late 19th Century. However, recent developments in material science and microelectronics allow us to envisage miniaturized systems combining a source of ambient mechanical energy and a sensor supplied by this source, possibly without any battery. Three families are generally distinguished for this conversion of mechanical energy into electricity, depending on the mode of transduction used: electromagnetic, piezoelectric and electrostatic. Each family has its advantages and drawbacks. In principle, we can say that electromagnetic transduction is the most effective in theory, but its performance drops as the device dimensions are scaled down. The piezoelectric transduction is efficient at all scales but requires constant stressing of an electroactive material, which raises reliability issues. Finally, the electrostatic transduction may be more complex to implement but is particularly suitable for miniaturized systems.

There are numerous applications for kinetic energy harvesting, even if there are few commercial products so far. For instance, mechanical structure health monitoring includes the monitoring of bridge oscillations, cracks in plane wings or changes in train rail fixtures with the aim to avoid deadly accidents. Other common examples include mechanical vibrations of the heart (can power an implanted pacemaker) or tire pressure monitoring systems (can be fed with mechanical energy of a rotating wheel).

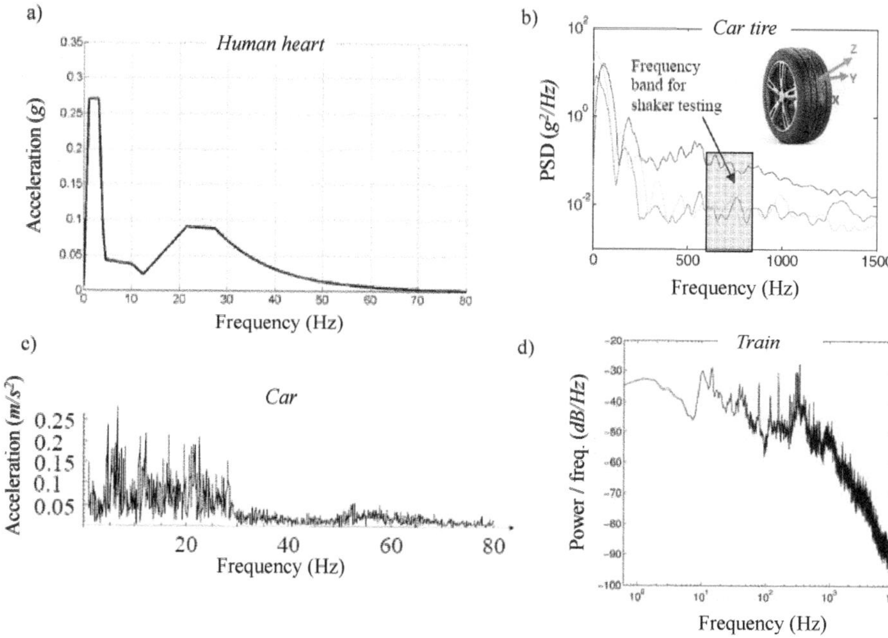

Figure I.2. *Examples of different vibration spectra published in the scientific literature: a) typical shape of the acceleration spectrum in the right atrium of a human heart [DET 11], b) power spectral density of the acceleration measured on the inner surface of a car tire driving at 60 km/h [REN 13], c) acceleration spectrum of a car [DES 05b] and d) acceleration of a train [VOC 14]*

Vibrations suitable for the generation of electricity can be of various forms. They can be periodic or non-periodic, harmonic or non-harmonic, regular or irregular, spread over a large frequency range or occurring at a single frequency. As shown in Figure I.2, each environment has a specific vibration frequency spectrum and it is very difficult, if not impossible, to design a generic KEH: each application needs a dedicated device in order to optimize the power yield.

1

Introduction to Electrostatic Kinetic Energy Harvesting

As a truly modern and complex system, the electrostatic kinetic energy harvester has several blocks, each performing its own function in the electrical or mechanical domains (Figure 1.1). We will briefly overview the role of the blocks in this introduction.

A large amount of energy can be produced by motion or vibrations in the mechanical domain. However it is not possible to use it directly in the electrical domain. Therefore, one would need a mechanism or a device that will transfer one form of energy (mechanical) to another (electrical). Such a device is called a *transducer*.

By definition, the transducer is a mechanism or a device that takes energy (power) in one form and converts it into another form. An electromechanical transducer is a very common and important type of transducers in modern microelectronics and microsystems: it converts mechanical energy into electrical form (and in some applications, vice versa, from the electrical to the mechanical domain). Thus, if one wishes to build a system that charges a storage capacitor in the electrical domain using the kinetic energy of some external motion in the mechanical domain, we should include an electromechanical transducer in such a system.

In the scheme of electrostatic kinetic energy harvesting shown in Figure 1.1, we use a variable capacitor as an electromechanical transducer. Let us consider a very simple case to understand its operation. Suppose that we have a parallel plate capacitor (Figure 1.2), with one of its two plates fixed and the other one movable. The distance

d between the plates is allowed to vary (due to any reason). The capacitance C_t of this capacitor depends on the distance d as follows:

$$C_t = \frac{\varepsilon_0 \varepsilon_r A}{d} \qquad [1.1]$$

where ε_0 is the permittivity of free space, ε_r is the relative permittivity of the medium between the plates (usually air and so $\varepsilon_d \approx 1$) and A is the area of plates.

If the capacitor is charged to Q_t, the energy of the electric field stored in it is

$$W_t = \frac{Q_t^2}{2C_t} \qquad [1.2]$$

Suppose now that we will manually move the movable plate farther from the fixed plate, keeping the charge Q_t constant. The new distance d_1 is greater than the original distance d: $d_1 > d$. The capacitance, according to equation [1.1], decreases and the stored energy, according to [1.2], increases.

In this example, we changed the energy stored in the electrical domain by manipulating a mechanical parameter (the distance d) of the transducer. This type of the electromechanical transducer is called *capacitive* (since it employs a capacitor) or *electrostatic*. There are other types of electromechanical transducers. Notably, electromagnetic and piezoelectric transducers are commonly used in kinetic energy harvesters, but they are out of the scope of this book.

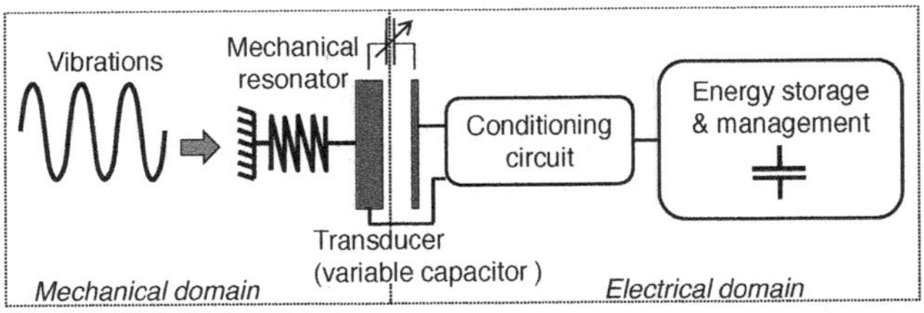

Figure 1.1. *Generic high-level structure of an electrostatic kinetic energy harvester (KEH) that includes a resonator, a capacitive transducer and a conditioning circuit (the transducer couples the mechanical and the electrical domain)*

We can reasonably assume that if we manage to attach one plate of such a capacitor to a vibrating or moving object, the external vibration will move the plate and change the distance d between the plates. This will affect the capacitance C_t and, in turn, will change the electrical energy stored in the capacitor. The work required to change electrical energy will be *provided by external mechanical forces* in the form of vibrations or any kind of motion that posseses kinetic energy. Hence, it gives the name to the system – kinetic or vibration energy harvesting. Note that the capacitor must be charged (or we can say that a voltage must be applied to it). Otherwise, without a charge or voltage, its electrical energy is zero.

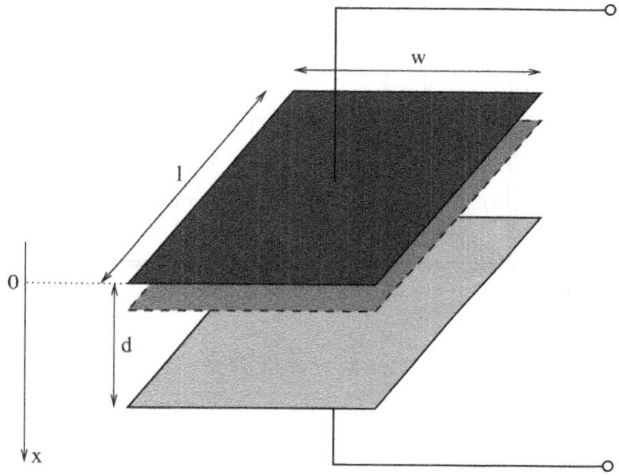

Figure 1.2. *Simplest electromechanical transducer in the form a variable parallel plates capacitor. The gap d of the capacitor can vary, which causes the variation of its capacitance*

Variable capacitors (and electromechanical transducers in general) are never attached directly to moving mechanical objects – this would be technologically impossible. There must be an intermediate system that will transfer external motion due to the environment to the motion of a capacitive transducer. This system will be a purely mechanical system as it involves the transformation of motion entirely in the mechanical domain.

How could we arrange a proper connection of a variable capacitor to a vibrating environment? Consider a system shown schematically in Figure 1.3. It has a simple parallel plate capacitor we just discussed, with one fixed electrode and one movable electrode. The movable electrode is suspended on a spring and also serves as a proof mass. The spring is attached to an enclosing frame. Now the frame is placed in the vibrating environment and experiences an acceleration due to external vibrations. Now, according to the laws of classical mechanics, the inertia of the mobile mass captures the energy from the external vibrations of the environment. The inertia generates the force $F_{ext}(t) = -mA_{ext}(t)$ in the mechanical domain where $A_{ext}(t)$ is the acceleration of the external vibrations. The inertia force causes the displacement of the proof mass with respect to the frame. Therefore, it also causes the mechanical displacement of the electrode of the capacitive transducer that is required for energy conversion.

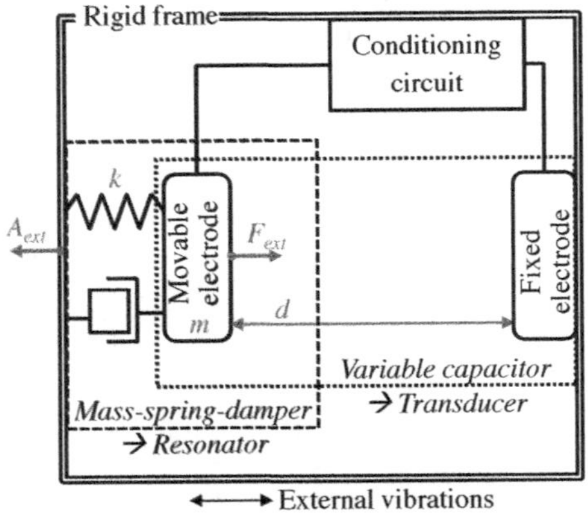

Figure 1.3. *Schematic view of a KEH including a resonator, a transducer and a conditioning circuit*

The mass and the spring form a *resonator*. There are always some losses due to dissipation (for example air damping) in a realistic mechanical system that we usually model as a damper. Therefore, we usually say that it is a mass–spring–damper system or a damped resonator. The aim of the resonator is to constrain the motion of the proof mass (electrode) and capture external vibrations in the most efficient way, usually through the phenomenon of *resonance*.

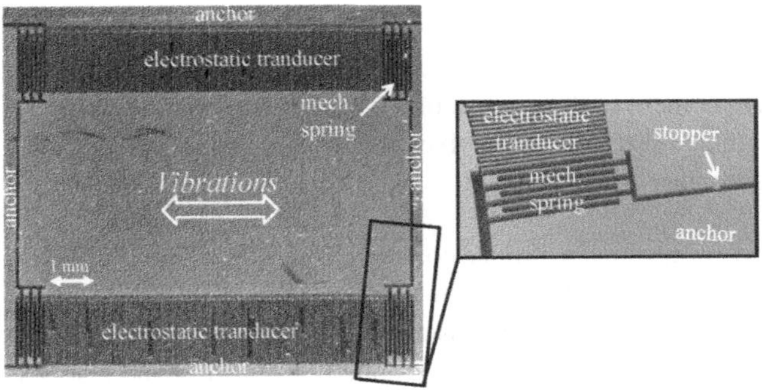

Figure 1.4. *Photo of a MEMS resonator and an electrostatic transducer for an e-KEH described in [BAS 14]*

Actual resonators do not look like the simplified structure as shown in Figure 1.3 with a proof mass on a spring. Realistic KEH resonators are distributed mechanical structures suspended on elastic arms, sometimes of a very complex form. One example is shown in Figure 1.4. To describe vibrations in a resonator very accurately and precisely, we could model it using a partial differential equation or even a finite element based method. However, the 1D model shown in Figure 1.3 is a very useful approximation, simple and convenient for the use in analytical and semianalytic studies of KEHs. In order to use the 1D model, we provide the mass m of the resonator (proof mass) and its equivalent spring coefficient k [GRA 12, NAY 05].

In Figure 1.3, we show that the conducting proof mass is, at the same time, a part of the resonator and a part of the transducer. Thus, as seen from the figure, the variable capacitor couples the mechanical and the electrical domains. We mentioned that the variable capacitor must be charged (or a voltage must be applied to it) in order to obtain useful operation. Knowing the purpose of the resonator and the transducer, let us briefly describe the conversion of energy from the mechanical to the electrical domain. If the proof mass (movable electrode) is displaced by dx due to external vibrations, this causes a change dC_t in the capacitance. The energy of the transducer changes according to equation [1.2]:

$$W_t + dW_t = \frac{Q_t^2}{2(C_t + dC_t)} \approx \frac{Q_t^2}{2C_t}\left(1 - \frac{dC_t}{C_t}\right) \qquad [1.3]$$

We will increase electrical energy (i.e. convert a portion of energy from the mechanical domain to the electrical domain) if $dW_t > 0$. Thus, dC_t must be negative, i.e. the capacitance C_t of the transducer must decrease. And vice versa, energy is converted from the electrical to the mechanical domain if $dW_t < 0$ and the capacitance increases, $dC_t > 0$. Therefore, energy conversion from the mechanical to the electrical domain occurs when the mobile electrode of a charged transducer moves in such a way that the capacitance of the transducer decreases [GAL 13b, DUD 14].

The transduction is obtained when an electrical force F_t generated by the transducer impedes the displacement of the resonator mass induced by external vibrations. For e-KEH, this force is the electrostatic force between the charged electrodes. The negative work of this force corresponds to the energy converted from the mechanical to the electrical domains. The generation of this force is controlled by the electrical state of the transducer (e.g. by the voltage across its terminals), which in turn is controlled by a *conditioning circuit*. The conditioning circuit operates entirely in the electrical domain. Later, the converted energy is managed, stored and supplied to the load.

In conclusion, in order to capture the motion of the environment, we use a mass–spring–damper system (resonator). The resonator drives the movable plate of a variable capacitor (transducer) changing its capacitance and therefore changing its electrical energy. A conditioning circuit then controls the voltage applied to the transducer (or its charge) in order to ensure useful energy conversion. Converted energy is further managed and stored. The resonator operates in the mechanical domain while the conditioning circuit, the power management circuit and energy storage subsystem operate in the electrical domain. The variable capacitor (transducer) couples the mechanical and electrical domains.

Energy harvesting principles that have been briefly presented in this introductory chapter will be analyzed thoroughly in the following chapters of the book. Chapter 2 discusses capacitive transducers and electromechanical conversion achieved by these devices. Chapters 3 and 4 provide a discussion on the mechanical component of a kinetic energy harvester. Chapters 5 and 6 are focused on nonlinear effects arising in KEHs and provide methods for the analysis of nonlinear systems. Chapter 7 discusses the fabrication technology of a microelectromechanical system (MEMS) KEH employing electrostatic transduction. Chapters 8 through 11 explain the role of the conditioning circuit and discuss the architecture of main families of conditioning circuits.

2

Capacitive Transducers

Widespread use of capacitive transducers has become possible because of the miniaturization of electronic systems. Indeed, capacitive transducers are inefficient at macroscale, so capacitive transducers are mainly implemented with silicon-based microelectromechanical system (MEMS) technologies. Capacitive transducers are used either as sensors or actuators for the transfer of information between the mechanical and electrical domains.

For information processing, the functions describing relations between mechanical and electrical quantities should be linear. For that reason, the preferable mode of operation of a transducer is generally a small-signal mode, where the magnitude of dynamic quantities is small enough so that the nonlinear distortions are negligible.

However, the energy conversion sets very different constraints. Not only is the linearity of the conversion unimportant, but also in many cases nonlinear behavior of electrical and mechanical devices is unavoidable or even desirable. And since the energy conversion generally operates in a *large amplitude* mode, the linearized small-signal mode is not adequate for their behavior modeling. As we will see, a capacitive transducer is an intrinsically nonlinear device. Moreover, in the mode of the energy conversion, the capacitive transducer is associated with a conditioning electronic, which is also nonlinear.

This chapter presents basic information describing capacitive transducers used as converter of mechanical energy into electricity.

2.1. Presentation of capacitive transducers

A capacitive transducer is a physical capacitor whose geometry can change in time so as to affect the value of the capacitance. Although a capacitor can be of any geometrical shape, in practice the most common is a parallel plate capacitor, whose geometry is given in Figure 2.1(a). Such a capacitor is constituted from a pair of parallel conductive planes (electrodes) spaced by some distance, called *gap*. The capacitance of such a device is:

$$C_t = \varepsilon_0 \varepsilon_r \frac{S}{d} \qquad [2.1]$$

where d is the distance between the planes (the gap), S is the overlapping area of the planes, ε_0 is the permittivity of vacuum (a fundamental constant equal to 8.85×10^{-12} F/m) and ε_r is the dielectric constant of the material between the electrodes.

It is very important to remember that [2.1] is only valid if the linear dimensions of the overlapping area are large compared to the gap d. In this case, the essential part of the flux of the electric field is formed by the homogeneous electric field concentrated between the plates. Equation [2.1] neglects the *fringing* electric field that, however, becomes dominant if d is large compared to the dimensions of the overlapping area (see discussion on the area-overlap transducer later in this section).

The capacitance of a parallel plate capacitor is a function of three parameters, and a variation of any of them produces a change in the capacitance:

$$C_t(t) = \varepsilon_0 \varepsilon_r(t) \frac{S(t)}{d(t)}, \qquad [2.2]$$

Most existing variable capacitors operate in air or in vacuum, so that $\varepsilon_r \approx 1$. However, there are exotic cases where the variation of the capacitance is produced by a motion of the dielectric material separating the electrodes (see Figure 2.1(b)), particularly in fluidic devices [BU 12]. A variable capacitor is usually obtained when one electrode of the capacitor moves with regard to the other. To simplify the analysis, it is usually considered that one electrode of the capacitor is fixed and the other moves. This is the most common configuration in energy harvesters, and will generally be assumed in this book, although there are many other applications of capacitive transducers where both electrodes are mobile [GAL 05].

In principle, the motion can be in any direction, but in the majority of capacitive transducers there are only two possible and exclusive kinds of motion: (1) electrodes move in their plane or (2) electrodes move along the axis normal to their planes. The choice of the motion mode is obtained by implementation of a particular geometry of capacitor, so that all undesirable directions of motion are blocked. Let us consider both cases:

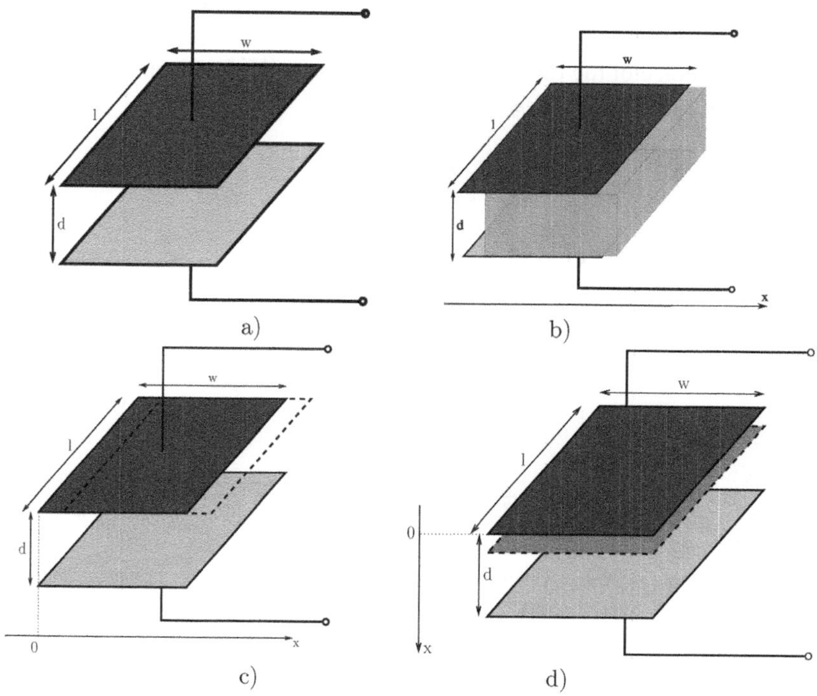

Figure 2.1. *Diagram presenting the geometry of capacitive transducer: a) Basic geometry of a parallel plate capacitor, b) geometry of a capacitive transducer with a movable dielectric, c) geometry of a transducer with parallel motion of electrodes and d) geometry of a capacitive transducer with gap-closing geometry*

1) *Parallel motion of electrodes*: In this case, the distance between the electrodes is kept constant, and the capacitance varies according to:

$$C_t(t) = \varepsilon_0 \frac{S(t)}{d} \qquad [2.3]$$

Such a capacitor is called an *area overlap capacitor* (Figure 2.1(c)). The variation of the overlapping area can be related to the relative displacement x of the electrodes by a function $S(x)$, where the function $S(\cdot)$ depends on the geometry of the transducer. If the transducer electrodes have a rectangular shape and the mobile electrode moves along one of its sides, the function S is given by:

$$S(x) = l(x_0 \pm x), \qquad [2.4]$$

where l is the length of the electrode side perpendicular to the motion and x_0 is the length of the overlapping rectangular area at rest. The parameter x_0 and the sign of x depend on the choice of the reference frame. For the structure given in Figure 2.1(c), $x_0 = w$, and $S(x)$ is expressed as

$$S(x) = l(w - |x|) \quad [2.5]$$

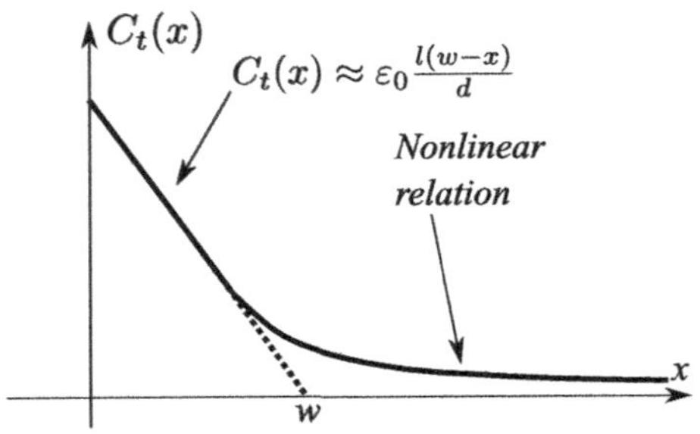

Figure 2.2. *Typical relationship between the capacitance of an area overlap transducer (Figure 2.1(c)) and the position of the movable electrode. For the case when the overlapping area is large, $C_t(x)$ is linear, otherwise, when the overlapping area goes to zero and becomes negative, the characteristic is nonlinear*

In general, the relationship between the capacitance of a parallel plate transducer and the position of the mobile electrode (x) can be expressed as

$$C_t(x) = \varepsilon_0 \frac{l(x_0 \pm x)}{d} \quad [2.6]$$

Formula [2.6] is only valid if $x_0 \pm x \gg d$. As we mentioned when equation [2.1] was introduced, the model of a parallel plate capacitor is only valid when the dimensions of the overlapping area are much greater than the gap. If the overlapping area goes to zero, equation [2.4] gives zero, but there is residual non-zero capacitance that cannot be neglected anymore.

A typical plot of capacitance for a rectangular area overlap transducer versus position of the mobile electrode is given in Figure 2.2. As can be seen, the relation between C_t and x is linear only as far as the overlapping area is large. Even if no overlap exists and the plates are separated far from each other (formally, it corresponds to a negative overlapping area), the capacitance is still not zero. Ignoring this point may lead to completely wrong results in analysis and simulation.

2) *Perpendicular motion of electrodes*: A transducer whose mobile electrode moves along the direction normal to the plane is called a *gap closing variable capacitor* (Figure 2.1(d)). Its capacitance changes according to:

$$C_t(t) = \varepsilon_0 \frac{S}{d(t)} \quad [2.7]$$

The transducer gap is linearly related to x:

$$d(t) = d_0 \pm x(t). \quad [2.8]$$

d_0 is the initial gap of transducer at rest ($x = 0$). d_0 and the sign before x depend on the choice of the reference frame.

The dependence of the capacitance on the displacement of electrodes is

$$C_t(x) = \varepsilon_0 \frac{S}{d_0 \pm x} \quad [2.9]$$

$C_t(x)$ tends toward infinity when the denominator becomes zero. In practice, this corresponds to a strong increase in the transducer capacitance when the electrodes become separated by a very small distance. In this case, the attracting force approaches infinity, resulting in a risk of instability (e.g. the pull-in phenomena described in Chapter 7). For this reason, stoppers are added in practical devices to prevent the reduction of gap between the electrodes below some critical value.

To conclude, it is important to keep in mind that the capacitance is a geometrical parameter and that the function $C_t(x)$ depends only on the geometry of the transducer.

2.2. Electrical operation of a variable capacitor

From an electrical point of view, a capacitor is a system of two electrodes separated by a dielectric or by vacuum. A capacitor behaves as an electrical element when its two electrodes have different electrical charges. If electrodes 1 and 2 have charges Q_1 and Q_2, the voltage between the electrodes is given by:

$$V_{12} = \phi_1 - \phi_2 = \frac{Q_1 - Q_2}{2C}. \quad [2.10]$$

where V_{12} is the voltage equal to a potential difference ($\phi_1 - \phi_2$) between the electrodes. The electrode having the higher charge has the higher potential.

In electronics, a capacitor is usually considered as an electrically neutral device, so that its electrodes have the same absolute charge, but of opposite sign. The neutrality of a capacitor is a very important hypothesis. In particular, it allows the application of Kirchhoff's laws for the circuit analysis. In practice, the neutrality of capacitors is ensured by existing DC (direct current) paths to the ground because of externally connected electrical devices or because of leakages.

The term *charge of the capacitor* designates the charge stored on one of the device's electrodes, chosen arbitrarily. This electrode is called *positive electrode* and labeled by the character "+" (see Figure 2.3). This figure demonstrates also the definition of the conventional positive voltage of the capacitor (the arrow points toward the positive electrode). The equation describing the capacitor is written as:

$$Q = CV. \tag{2.11}$$

The current–voltage characteristic of the transducer is obtained by derivation of the last expression:

$$i = \frac{\mathrm{d}Q}{\mathrm{d}t} = \frac{\mathrm{d}}{\mathrm{d}t}(CV) = C\frac{\mathrm{d}V}{\mathrm{d}t} + V\frac{\mathrm{d}C}{\mathrm{d}t}. \tag{2.12}$$

The two components of the current can be interpreted as "electrical component", proportional to the voltage time derivative, and the "motional component" proportional to the capacitance time derivative.

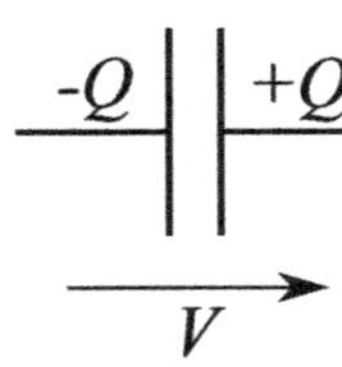

Figure 2.3. *Definition of the conventional sign of charge and voltage of a capacitor*

2.3. Energy and force in capacitive transducers

2.3.1. *Energy of a capacitor*

It can be shown that a capacitor stores energy. This is the potential energy of the electric field that exists between the electrodes of the capacitor. This field is generated by the charges of opposite signs distributed on the electrodes. From electrostatics, we know that the electrical potential energy, stored in this system, is

$$W = \frac{1}{2} \iint\limits_{\text{electrode surface}} \sigma(s)V(s)ds \qquad [2.13]$$

where σ is the surface charge density of the charge distributed on one electrode and V is the voltage between the electrodes. Since the plates are made up of a conductive material, they are equipotential. Therefore, their charges are concentrated on the surface. For this reason, we integrate expression [2.13] over the surface, but not the volume of the electrodes. This means that the voltage between any two points (one taken at the positively charged plate and the other – at the negatively charged plate) is the same. We can rewrite:

$$W = \frac{V}{2} \iint\limits_{\text{plate surface}} \sigma(s)ds = \frac{QV}{2} \qquad [2.14]$$

where Q is the total charge at the plate.

Note that formula [2.14] contains both, the charge and voltage. In the case of a capacitor, we cannot say that the charge and voltage are independent of each other. We can either assume that the voltage between the plates of the capacitor is caused by the charge that distributed on its plates, i.e. $V = V(Q, x)$ where x is the position of the plate. In this case, we consider that the charge Q is our "primary" variable that describes the electrical state of the capacitor. Or, we can assume that the voltage across a capacitor V defines the charge distributed on its plates. The latter occurs when we connect the capacitor to a battery: the voltage due to the battery dictates the charge in the capacitor. Thus, we say $Q = Q(V, x)$. But in this case, the system is composed of the capacitor *and* of the battery as an *external* source of voltage. As a result, when analyzing the behavior of the capacitor whose voltage is fixed, this should be taken into account.

With the use of the formula relating Q, V and C, the energy can also be expressed through the charge:

$$W = \frac{Q^2}{2C} \qquad [2.15]$$

or through the voltage:

$$W = \frac{CV^2}{2}. \qquad [2.16]$$

The latter formulas show that the potential energy of a capacitor is always positive. Another important point is the energy of the capacitor is a function of state (in the thermodynamic sense), which means that it depends only on the present electrical and mechanical state of the capacitor (charge, voltage, capacitance value), and not on the history of the system.

2.3.2. *Force of the capacitor*

The existence of mechanical effect (mechanical forces) caused by the internal electrical field of a capacitor allows the use of a variable capacitor as an electromechanical transducer. Mechanical forces are applied to all parts of a capacitor[1]. In this section, we discuss the derivation of the general formula for the force in a capacitor, and the fundamental relation existing between this force and the energy stored by a capacitor.

To calculate the mechanical force applied to a part of the capacitor along an axis x, the following thought experiment can be made:

– the considered part of the system should be made freely movable along the axis x;

– the considered part should be moved along this axis by infinitesimal distance $\mathrm{d}x$ (this is called "virtual displacement" in analytical mechanics);

– the capacitance variation $\mathrm{d}C_t$ should be measured.

The force F along the axis x is calculated as:

$$F = \frac{1}{2}V^2\frac{\mathrm{d}C_t}{\mathrm{d}x}, \qquad [2.17]$$

where V is the voltage on the capacitor.

As a result, an electrostatic force is applied to all parts of the capacitor whose geometric position impacts the value of the capacitance, namely the electrodes and the dielectric between the plates. The force is oriented in the direction of *increasing* the capacitance.

What is the origin of the force? We propose first to discuss the origin of the force applied to the capacitor electrodes. By applying a voltage between two conducting

1 A force is a notion from mechanics, but sometimes in the literature the forces created by electrical phenomena are called "electrical forces". Their action on mechanical systems are described by usual laws of mechanics.

plates, we cause the separation of charges: one plate will accumulate an excessive positive charge $+Q$, while the other plate will accumulate an excessive negative charge $-Q$. The attractive force is Coulomb's force due to the electrostatic interaction between the charges on the negatively charged plate and the charges on the positively charged plate. Since the two plates always have charges of the opposite polarities (due to an applied voltage), the resulting Coulomb's force is always attractive, as shown in Figure 2.4 (also see how the formula for the force [2.17] contains the squared voltage).

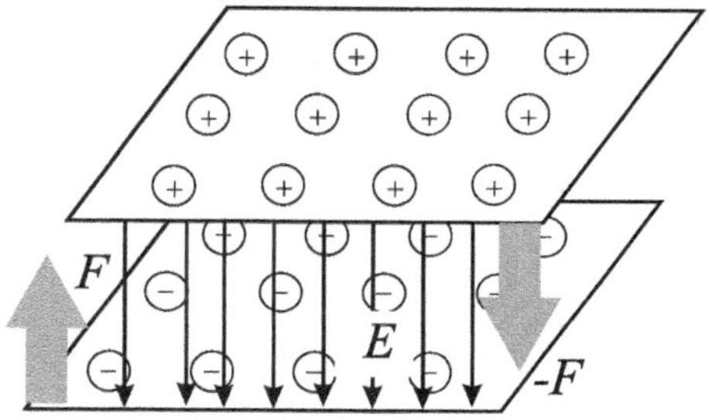

Figure 2.4. *The electric field and the attracting electrostatic force that acts between two conductors*

In order to derive the expression for F, we will have to take into account all Coulomb's forces that act on a selected elementary charge dq located at the position $\mathbf{r}$:

$$d\mathbf{F}(\mathbf{r}) = \frac{1}{4\pi\varepsilon_0} \sum_{\text{all charges}} \frac{dq \cdot dq'}{|\mathbf{r} - \mathbf{r}'|^2} \hat{r} \qquad [2.18]$$

where dq' are all other charges (on both plates) with the positions given by the radius vectors $\mathbf{r}'$ and $\hat{r}$ are the corresponding unit vectors. In order to find the total force that acts on the entire plate (positive or negative), we have to sum the individual forces $d\mathbf{F}$ for all charges that belong to that plate. The evaluation of this sum (or integral) is very complex. However, there is a simple approach that would allow us to derive the force using the concept of the capacitance.

The capacitor's force is a *potential* force, since it is related with the static electric field, which is known to be potential. We can use the relation between the potential energy of a conservative system and the potential force. In the one-dimensional case, if we consider an isolated capacitor, the force is

$$F = -\frac{dW}{dx} = -\frac{1}{2}\frac{d}{dx}\left(\frac{Q^2}{2C}\right) = \frac{1}{2}\frac{Q^2}{C^2}\frac{dC}{dx} = \frac{V^2}{2}\frac{dC}{dx} \qquad [2.19]$$

In the most general, three-dimensional case, we write

$$F = -\text{grad } W = -\frac{Q^2}{2}\text{grad } \frac{1}{C} = \frac{Q^2}{2C^2}\text{grad } C = \frac{V^2}{2}\text{grad } C. \qquad [2.20]$$

It also possible to find the force by using the relation between the potential energy and the voltage [2.16], assuming that we fix a constant voltage V across the capacitor and so $Q = Q(V, x)$. However, as we mentioned in section 2.3.1, we must account for the presence of the voltage source. Indeed, during the virtual displacement $\mathrm{d}x$, the battery does work on the plates. The equation that defines the infinitesimal change in energy of the capacitor is

$$dW = -Fdx + VdQ \qquad [2.21]$$

where VdQ is the term that represent the work done by the battery that provides V. And so we obtain

$$F = -\frac{dW}{dx} + V\frac{dQ}{dx} = -\frac{V^2}{2}\frac{dC}{dx} + V\frac{d(CV)}{dx} = \frac{V^2}{2}\frac{dC}{dx} \qquad [2.22]$$

Without any surprise, we obtain exactly the same expression as when the capacitor was considered isolated, with fixed charge. The electrostatic force acting on the capacitor is *entirely and exclusively* defined by the distribution and amount of charge on its plates, and similarly with the energy, it is a *function of the state.*

The role of this force in electromechanical conversion and different aspects of its influence on the resonator are discussed in Chapters 3–6 dealt with the operation of KEH in the mechanical domain.

We now present the expressions of the force specific for the geometries of transducer discussed in the previous section. For an area overlap capacitor biased by a voltage V_t, whose capacitance is given by [2.6], the transducer force F_t is independent of the displacement:

$$F_t = \pm\frac{1}{2}V_t^2\varepsilon_0\frac{l}{d} \qquad [2.23]$$

For a gap closing transistor whose capacitance is given by [2.9], the transducer's force is:

$$F_t = \mp\frac{1}{2}V_t^2\varepsilon_0\frac{S}{(d_0 \pm x)^2} \qquad [2.24]$$

For the case where the transducer gap goes to zero, the force becomes infinite.

These expressions explain why a capacitive transducer is only efficient at the microscale. Indeed, the transducer force does not scale with the device dimensions, since both S and $(d_0 - x)^2$ scale quadratically, whereas mechanical forces are proportional to the linear dimensions of the spring and to the cube for the mass inertia. This means that electrostatic forces are too weak to be useful at the macroscale.

These expressions also highlight the dependence of the transducer force on the voltage. The ability to modulate the transducer force by the applied voltage is a powerful tool for electrical synthesis of mechanical behavior. However, the transducer force is unilateral, i.e. cannot change its sign because of the square dependence of the voltage. This creates some problems both for sensor/actuator implementations and for energy conversion. One possible solution is the use of a differential capacitive transducer. However, while differential structures are often used for sensing/actuating applications, they are rarely used in energy harvesting devices.

2.3.3. *Capacitive transducers biased by an electret layer*

The capacitive transducers presented so far are passive devices: they need to be biased in order to generate electricity. For this reason, sometimes a layer of dielectric with embedded electrical charges called *electret* is added to the transducer. This section deals with general presentation of electrostatic transducers with electret and with the establishment of its electrical model. More information about technological and material aspects of electret biasing can be found in Chapter 7, and its application in e-KEH is discussed in Chapters 8 and 11.

A capacitive transducer with electret consists of a variable capacitor with an additional thin dielectric layer, of thickness and dielectric constant d_{el} and ϵ_{el}, respectively, attached to one of the electrodes and embedding a constant charge Q_{el}, as shown in Figure 2.5(a) [SUZ 11].

As we said earlier, in electrical analysis it is usually considered that the devices are electrically neutral. This hypothesis of neutrality is essential for analysis of the capacitive transducer with electret. According to it, the charges of the electrodes Q_1, Q_2 and the charge of electret Q_{el} are bounded by the relation:

$$Q_1 + Q_2 + Q_{el} = 0 \quad [2.25]$$

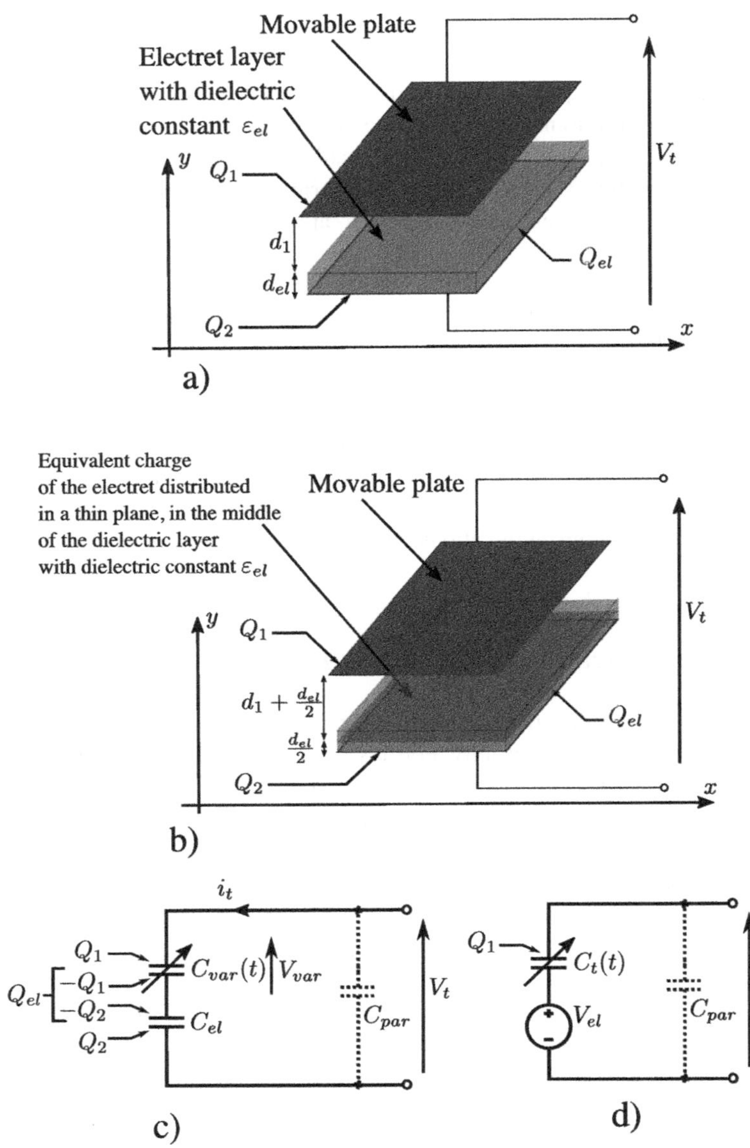

Figure 2.5. *a) Physical representation of an electret e-KEH; b) equivalent physical model in which the charges of electret are focused in a thick plane; c) detailed equivalent electrical model and d) compact equivalent electrical model*

In this way, any change in Q_1 causes a change in Q_2. It is easy to see that if this relation is not assumed, the currents in the two terminals would be different: such a possibility is not admitted by the circuit theory for a dipole.

From basic electricity course, it is known that an infinite layer of uniformly distributed charges with thickness d_{el} creates outside of it exactly the same electrical field as an infinitely thin layer (a plane) having the same surface charge density, placed in the middle of the layer. We propose to use this equivalence for the derivation of the current–voltage characteristic of the transducer.

The equivalent diagram of a capacitive transducer with infinitely thin electret layer is given in Figure 2.5(b). For the analysis, we suppose that this plane with the charge Q_{el} is a superposition of two planes with charges $-Q_1$ and $-Q_2$, according to [2.25]. Moreover, the basic electricity laws allow us to suppose that these two planes are conductive. We obtain a structure composed of two capacitors with capacitances $C_{var} = \epsilon_0 S/(d_1 + d_{el}/(2\varepsilon_{el}))$ and $C_{el} = 2\varepsilon_{el}\epsilon_0 S/d_{el}$ in series, so that we have for C_t:

$$C_t = \frac{C_{var}C_{el}}{C_{var} + C_{el}}, \qquad [2.26]$$

where d_1 is the thickness of the air gap in the transducer, $d_1 = d_0 - d_{el}$.

To determine the expression of V_t we can write:

$$\begin{aligned} &V_t(t) = V_{var} - V_{el} = Q_1/C_{var} - Q_2/C_{el} = \\ &Q_1(1/C_{var} + 1/C_{el}) + Q_{el}/C_{el} = \tfrac{Q_1}{C_t} + V_{el}. \end{aligned} \qquad [2.27]$$

where $V_{el} = Q_{el}/C_{el}$ is the *built-in* voltage of the electret. The electret e-KEH can now be represented just by a dipole made of the variable capacitance C_t in series with the DC voltage source V_{el} (Figure 2.5(d)).

The relationship between the voltage and the current in the transducer is given by:

$$i_t = \frac{\mathrm{d}Q_1}{\mathrm{d}t} = \frac{\mathrm{d}}{\mathrm{d}t}[(V_t + V_{el})C_t]. \qquad [2.28]$$

This relation highlight a fundamental difference with a passive capacitive transducer. Suppose that both passive and electret-biased transducers are short circuited and their capacitance C_t varies in time. In the passive transducer, there will be no electrical current, whereas the device with electret generates a motional current given by:

$$i_t = \frac{dQ_1}{dt} = V_{el}\frac{dC_t}{dt}. \qquad [2.29]$$

Such a system, when loaded with a resistance, generates electrical power by converting energy from the mechanical into the electrical domain. More details about the conditioning electronics used with transducers biased by electret will be given in Chapters 8 and 10.

The equivalent schematics of transducer (Figures 2.5(c) and (d)) include the parasitic capacitance of the structure, which is connected in parallel with the transducer's electrodes. This capacitance is due to the transducer geometry (see Chapter 7) and to the connection wires. If C_{par} is comparable to the minimum value of the variable capacitance C_t, it may have a substantial impact on the device operation, and must be accounted for in the model.

The proposed electrical model of the capacitive transducer with electret is valid both for the gap-closing motion along the axis y and for the area-overlapping motion along the axis x. The following question may be asked: is a transducer with the electret a different device from a passive capacitive transducer? From a standpoint of the electrical engineering, our answer is clearly not: the only difference between them is that the passive transducer needs an external biasing, whereas a transducer with electret has an embedded biasing. The equivalent electrical circuit of a transducer with electret separates clearly the capacitive transduction from the biasing. In Chapter 8, it will be shown that similar conditioning circuits can often be used for both kinds of transducers.

2.4. Energy conversion with a capacitive transducer

The work of the force generated by the capacitive transducer represents the energy transferred between the mechanical and the electrical domains. Suppose that during a time interval $[t_1, t_2]$ the capacitance C_t of a transducer changes monotonically. The work produced by the force of the transducer is given by:

$$W_t = \int_{t_1}^{t_2} F_t v dt = \int_{t_1}^{t_2} \frac{1}{2}V_t^2 \frac{dC_t}{dx}\frac{dx}{dt}dt = \int_{C_1}^{C_2} \frac{1}{2}V_t^2 dC_t, \qquad [2.30]$$

where $C_1 = C_t(t_1)$, $C_2 = C_t(t_2)$, v is the velocity of the mobile electrode.

When this work is positive, the transducer force F_t increases the energy in the mechanical domain (namely, the velocity of the mobile electrode), and at the same time the electrical energy decreases. When this work is negative, there is a transfer of energy from mechanical domain to electrical domain, which is used in the energy conversion.

Since $V_t^2 \geq 0$, the work is positive when the capacitance increases and is negative when the capacitance decreases. As a result, *the energy transfer from the mechanical domain to electrical domain occurs when a biased capacitive transducer decreases its capacitance*. For the energy harvesting applications, the voltage on the transducer should be minimized when the capacitance increases and maximized when the capacitance decreases.

The quantity C_t evolving in time can be seen as a path defined in the plane (V_t, Q_t), given by the relation $C_t = Q_t/V_t$, and the energy can be seen as the path integral. This allows an application of the formalism of vector calculus. Given $\mathrm{d}C_t = \mathrm{d}Q_t/V_t - (Q_t/V_t^2)\mathrm{d}V_t$, the equation [2.30] can be written as:

$$W_t = \frac{1}{2}\int_\Gamma \left[V_t \mathrm{d}Q_t - Q_t \mathrm{d}V_t\right], \qquad [2.31]$$

where Γ is the path that the transducer state follows in the plane QV between the times t_1 and t_2, and $\int_\Gamma$ is the path integral.

If the curve Γ forms a cycle, the work W is written as

$$W_t = \frac{1}{2}\oint_\Gamma \left[V_t \mathrm{d}Q_t - Q_t \mathrm{d}V_t\right]. \qquad [2.32]$$

According to Green's theorem, this formula expresses the area enclosed by the curve Γ, if the path is negatively oriented in the plane (V_t, Q_t), i.e. the path is counter clockwise in Figure 2.6 (see the proof in the appendix at the end of this chapter (section 2.8)). In this case, equation [2.32] is negative, and the area enclosed by the path represents the energy converted into electricity during the cycle of C_t variation. The line representing the state of the transducer in the (V_t, Q_t) plane is called *QV diagram*, and is a very elegant representation widely used for the analysis of the energy conversion achieved by capacitive transducers.

2.5. Optimization of the operation of a capacitive transducer

The goal when designing electrostatic energy harvesters is to maximize the area of the *QV* cycle corresponding to the converted energy. Let us consider a transducer whose capacitance variation is between C_{min} and C_{max}. In the QV plane, all possible states of this transducer are limited by the loci $Q_t = C_{min}V_t$ and $Q_t = C_{max}V_t$ (Figure 2.7). They define an open segment with infinite area on the QV plane, if there is no limit on the voltage on the transducer. In practice, the voltage is always limited by the technology, let us say, by a value V_{max}. In this way, an ideal QV cycle is a triangle, formed by the lines $C_t = C_{max}$, $C_t = C_{min}$, $V_t = V_{max}$ (triangle OMN, Figure 2.7). Its area is given by:

$$\Delta W = \frac{1}{2}V_{max}^2(C_{max} - C_{min}). \quad [2.33]$$

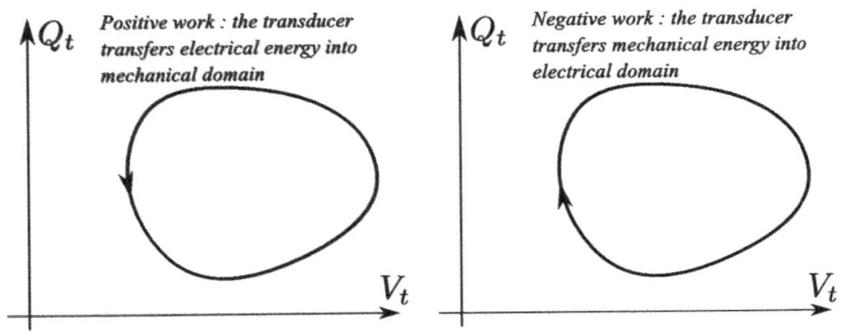

Figure 2.6. *Example of cyclic operation of capacitive transducer plotted in the charge–voltage plane*

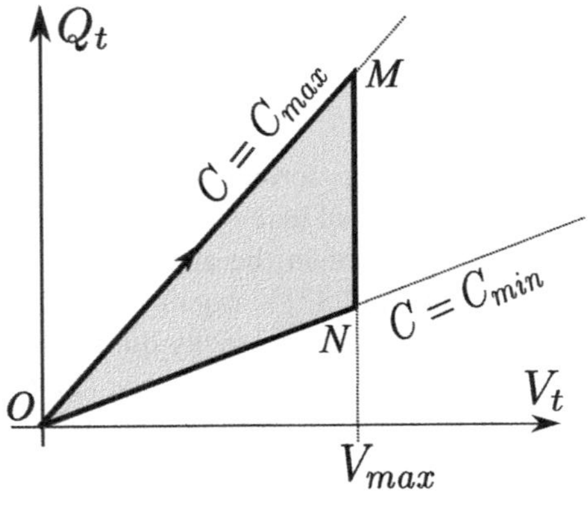

Figure 2.7. *QV cycle maximizing the converted energy at one cycle for a transducer capacitance varying from C_{min} to C_{max} and under constraint of a maximum voltage of V_{max}*

Such a QV cycle is called a constant voltage QV cycle [MEN 01]. This term emphasizes the fact that the energy conversion is achieved when the voltage on the transducer is constant.

This formula provides an opportunity to estimate the maximum energy and power that can be generated by a capacitive transducer in a realistic context. If we take the value for C_{min} and C_{max} from a 1 cm^2 MEMS capacitive transducer [BAS 14] (40 and 140 pF, respectively) and 50 V for V_{max} (the limit for the 0.35 μm technology of AMS), we obtain 125 nJ per cycle. And with the frequency of the capacitance variation at 100 Hz, it corresponds to 12 μW of converted power. This figure should be seen as the order of magnitude of maximum convertible power with capacitive transducers at microscale, even if this value can be increased if we assume different hypotheses on the frequency, the transducer parameters and the maximal voltages.

As shown by equation [2.30], the operation of a capacitive transducer is fully controlled by the voltage waveform $V_t(t)$ applied across its electrodes while the transducer capacitance C_t varies. Indeed, for each value of C_t, as far as V_t is defined, the electrical state of the transducer is uniquely defined through the formula $Q_t = C_t V_t$. In this way, by generating an appropriate voltage waveform on the transducer, it is possible to "synthesize" any desirable QV cycle. This is one of the roles of the conditioning circuit: the definition of a dynamic biasing required for the energy conversion by the transducer.

However, the QV cycles implemented practically are often different from the optimal cycle given in Figure 2.7. The first reason is the difficulty to generate the optimal QV cycle at a reasonable energy cost. Second, the optimization of the converted energy is only one of the two roles of the conditioning circuit. The other role is the optimal transmission of the converted energy toward the storage or load device. The compromise between the efficiency of these two functions results in suboptimal power conversion. The analysis and design of conditioning circuits will be discussed in Chapters 8–11.

2.6. Electromechanical coupling

In the earlier sections, we assumed a defined variation of the transducer capacitance, between C_{min} and C_{max}. This hypothesis is non-realistic and can be highlighted in the following mental experiment. Imagine a transducer attached to a given resonator submitted to external vibrations. For some conditions, the capacitance of the transducer varies between C_{min} and C_{max}. As will be shown in section 3.2, there is an upper bound of the power $P_{in,max}$ that the system is able to absorb from the external vibrations, and hence that it is able to deliver to the electrical domain. Suppose that the triangular cycle as shown in Figure 2.7 is used. From the other side, according to equation [2.33], the energy converted by the transducer can have any large value, if the voltage V_{max} is not upper bounded. There is an apparent contradiction, which is solved by the consideration of the electromechanical coupling. Indeed, assuming a given variation of the transducer capacitance is equivalent to assume a given motion of the mobile mass. However, the energy conversion is carried out through an application to the mass of the transducer's force. This force is proportional to the square of the voltage. If the voltage is high, the force is large, and the motion of the mobile mass is likely to be impeded by the process of the energy conversion. As a result, the capacitance variation of the transducer is reduced, therefore enforcing the fundamental limit given by $P_{in,max}$.

This situation explains the difficulty in analysis and design of capacitive vibration energy harvesters. In order to analyze the energy conversion of the transducer, the capacitance variation (and hence, the motion of the mobile mass) should be known, but the mechanical dynamics of the system are strongly affected by the electrical operation of the transducer, especially when the energy conversion is to be maximized. More insight into the methods allowing analysis and design of kinetic energy harvesters with capacitive conversion will be presented in Chapters 5 and 6.

2.7. Conclusions

This section presented an overview of different aspects related to the capacitive transduction. Many of these aspects will be discussed in detail in the following chapters.

2.8. Appendix: proof of formula [2.32] for the energy converted in a cycle

In this section, we present the demonstration of the fact that the area of a charge–voltage cycle performed by a variable capacitance is numerically equal to the electrical energy generated or absorbed by the capacitance, depending on the cycle direction.

The demonstration starts from formula [2.32] expressing the work achieved by the capacitive transducer in the mechanical domain:

$$W_t = \frac{1}{2} \oint_\Gamma \left[V_t \mathrm{d}Q_t - Q_t \mathrm{d}V_t \right]. \qquad [2.34]$$

The Green theorem states that for a positively oriented, piecewise smooth, simple closed curve Γ in a plane, the region D bounded by Γ and functions L and M of (x, y)

defined on an open region containing D and having continuous partial derivatives, the following equality is true [RIL 06]:

$$\oint_{\Gamma}(Ldx + Mdy) = \iint\limits_{D} \left[\frac{\partial M}{\partial x} - \frac{\partial L}{\partial y}\right] \mathrm{d}x\mathrm{d}y. \qquad [2.35]$$

Applying this theorem to equation [2.34], we get:

$$\begin{aligned} W_t &= \frac{1}{2}\oint_{\Gamma}[V_t \mathrm{d}Q_t - Q_t \mathrm{d}V_t] \\ &= \frac{1}{2}\iint\limits_{D}[-1-1]\mathrm{d}Q_t \mathrm{d}V_t = -\iint\limits_{D} \mathrm{d}Q_t \mathrm{d}V_t. \end{aligned} \qquad [2.36]$$

The last double integral express the area of the domain D enclosed by the curve. The minus sign means that the work is positive if the path orientation is negative (counterclockwise). For a positively oriented path, the work is negative, and the energy is converted from the mechanical domain into the electrical domain.

3

Mechanical Aspects of Kinetic Energy Harvesters: Linear Resonators

The aim of this chapter is to explain the role and operation of the mechanical part of kinetic energy harvesters (KEHs). We introduce fundamental concepts such as linear resonators, resonance and power balance in KEHs. We discuss the model of free and harmonically driven linear resonators. In particular, we show that the solution of a forced linear resonator is a harmonic oscillation at the frequency of the external signal.

3.1. Overview of mechanical forces and the resonator model

3.1.1. *Linear resonator as the main model of the mechanical part*

We already know that the resonator in a KEH is formed by a movable electrode of mass m (proof mass) that is suspended on a spring with a spring factor k, see Figure 3.1. In order to account for damping due to viscous friction with air, we add a dashpot (damper) with a coefficient μ into our model. The force that drives the resonator is $F_{ext}(t)$. It is the inertial force of the frame caused by external vibrations (see the discussion in Chapter 1). We will denote the displacement of the proof mass relative to the frame as $x(t)$.

In order to write the equation that describes the relative displacement $x(t)$, we will apply the second Newtonian Law: the superposition of all forces acting on the proof mass is zero:

$$m\ddot{x} = F_{ext} + F_{damping} + F_{spring} \qquad [3.1]$$

where $F_{damping} = -\mu v$, $F_{spring} = -kx$, and $v = \dot{x}$ is the velocity of the proof mass. Note that the overdot denotes the derivative with respect to time. Everywhere in this chapter, we assume that $F_{ext} = F_{ext}(t)$, $x = x(t)$ and $v = v(t)$. We will omit the notation of the dependence on time for simplicity. Taking this into account, we write equation [3.1] in its conventional form, known as the mass-spring-damper equation:

$$m\ddot{x} + \mu\dot{x} + kx = F_{ext} \qquad [3.2]$$

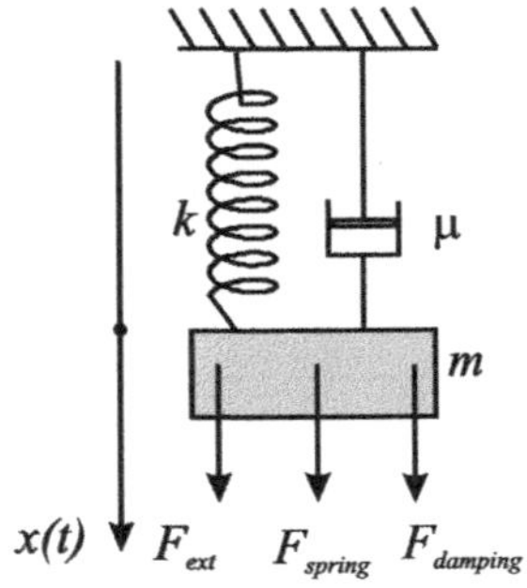

Figure 3.1. *Schematic view of a resonator driven by an external (mechanical) force* F_{ext}

Equation [3.2] is one of most commonly used in the analysis of KEH resonators and is extremely common in research literature. We will predominantly use this equation throughout the book.

The external force, in the most general case, can be of any form. However, we will consider the case when

$$F_{ext} = \tilde{F}_0 \cos(\omega_{ext}t + \phi_0) = mA_{ext} \cos(\omega_{ext}t + \phi_0) \qquad [3.3]$$

where A_{ext} is the amplitude of the external vibrations, ω_{ext} is its frequency and ϕ_0 is an initial phase.

Note that the form of the force with an arbitrary initial phase ϕ_0 is very convenient and also very general. For example, if $\phi_0 = \pi/2$, $\cos(\omega_{ext}t + \pi/2) = -\sin(\omega_{ext}t)$ and this form of the input (external) vibrations is commonly used in the literature on kinetic energy harvesting. In addition, the phase can be eliminated from the equation without affecting its solution. Indeed, the simple time transform $t \rightarrow t' - \phi_0/\omega_{ext}$, where t' is new time, removes the phase from the right-hand side of equation [3.3] (but does not affect the derivatives in the left-hand side).

Another case that we will briefly discuss in this chapter are multi-harmonic vibrations that contain a set of harmonic components:

$$F_{ext} = \sum_{n=1}^{N} \tilde{F}_{0,n} \cos(\omega_{ext,n} t + \phi_{0,n}) = \sum_{n=1}^{N} m A_{ext,n} \cos(\omega_{ext,n} t + \phi_{0,n}) \quad [3.4]$$

where $\tilde{F}_{0,n}$ are the amplitudes of the harmonic components, $\omega_{ext,n}$ are their frequencies and $\phi_{0,n}$ are their phases.

Expression [3.4] also describes the case of a periodic but not necessarily harmonic force. Indeed, here we can recall that any periodic signal, including a force, can be presented using a (truncated) Fourier series that also has the form of equation [3.4]. However, in the case of the Fourier series, we have to state that $\omega_{ext,n} = n \cdot 2\pi/T$, where T is the period of the periodic external force. We also note that in this case, the amplitudes are $A_{ext,n} = \sqrt{a_n^2 + b_n^2}$ and the phases are $\tan(\phi_{0,n}) = b_n/a_n$, where a_n and b_n are the standard Fourier coefficients:

$$\begin{aligned} a_n &= \tfrac{2}{T} \int_0^T F_{ext}(t) \cos(n \cdot 2\pi t/T) dt \\ b_n &= \tfrac{2}{T} \int_0^T F_{ext}(t) \sin(n \cdot 2\pi t/T) dt \end{aligned} \quad [3.5]$$

Another form of equation [3.2] is obtained by dividing both sides of the equation by the mass m:

$$\ddot{x} + 2\gamma \dot{x} + \omega_0^2 x = F_{ext}/m \quad [3.6]$$

where $\gamma = \mu/(2m)$ is the dissipation (damping, friction) parameter and $\omega_0 = \sqrt{k/m}$ is the natural frequency of the resonator. This ordinary differential equation [3.6] is known as a *linear resonator*. In particular, in the case $\mu > 0$ or $\gamma > 0$, we say that it is a damped linear resonator or the linear resonator with dissipation, while in the case $\mu = 0$ we say it is a conservative or lossless resonator.

There are two particular cases of the resonator with dissipation: "small" dissipation $\gamma < \omega_0$ and "large" dissipation $\gamma > \omega_0$. The parameter that is very often used to distinguish between small and large dissipation is called the *quality factor* or the *Q*-factor. Fundamentally, it is given by the relation:

$$Q = 2\pi \frac{\text{Maximal energy stored in the system over a cycle}}{\text{Energy dissipated per one cycle}} = \omega \frac{\text{Energy stored}}{\text{Power loss}} \quad [3.7]$$

In general, the quality factor is a frequency-dependent parameter. In the case of a second-order resonator, it is given by:

$$Q = \frac{\omega_0}{2\gamma} \qquad [3.8]$$

In the context of formula [3.8], it is said that the case $Q > 1/2$ corresponds to an underdamped resonator while the case $Q < 1/2$ corresponds to an overdamped resonator.

In the case of a mechanical resonator, the Q-factor characterizes the losses in the resonator, which may be unwanted (for instance, due to friction) or intentionally introduced (for instance, in the case of energy harvesting systems, we use the transducer to remove the energy of the resonator and transfer it to the electrical domain). A resonator with a higher Q-factor will lose less energy during one period of natural oscillations $T_0 = 2\pi/\omega_0$ than another resonator with a smaller Q-factor during the same time T_0.

3.1.2. *The nature and effect of the transducer force*

The mechanical equation of the resonator, equation [3.2], describes the motion of the KEH resonator taking into account all mechanical forces and external vibrations that drive the resonator. However, we have not taken into account one more force, a force that is generated by the transducer – the transducer force F_t, discussed in the Chapter 2.

Recall that the proof mass, on one hand, is a part of the resonator, and, on the other hand, is also a part of the variable capacitor (transducer) since it is attached to the mobile electrode of the variable capacitor. It is known that if we apply a voltage between two conducting plates, there will appear an attractive electrostatic force. This means that as soon as we connect a conditioning circuit that controls the voltage across the transducer to it, we will introduce an additional force acting on the proof mass.

The transducer is, as a matter of fact, a capacitor that consists of two oppositely charged plates. A force will be introduced to the resonator as soon as we bias (apply voltage) the variable capacitor, and therefore we state that the transducer force F_t is given by [2.17]:

$$F_t = \frac{V_t^2}{2}\frac{dC_t}{dx} \qquad [3.9]$$

Here, V_t is the voltage across the capacitor and $C_t(x)$ is its capacitance–displacement characteristic. The nature of the force is purely electrical and it is caused by the separation of charges between the plates. However, the force has a mechanical effect: while acting on charges and trying to pull the two charged plates together, it initiates motion in the resonator/transducer.

As a result, if we want to take into account all forces that act on the resonator, we will write the mass-spring-damper equation [3.2] taking into account the transducer force:

$$m\ddot{x} + \mu\dot{x} + kx = F_{ext} + F_t \quad [3.10]$$

or, using the other form of this equation,

$$\ddot{x} + 2\gamma\dot{x} + \omega_0^2 x = F_{ext}/m + F_t/m \quad [3.11]$$

The inclusion of the transducer force F_t usually makes equation [3.11] nonlinear: in the most general case, the transducer force is nonlinear and, in many cases, it is piecewise defined. Nonlinear systems such as [3.11] introduce a range of complex behavior: electrostatic frequency softening or hardening, bifurcations and hysteresis. We will discuss these effects later in this book. In this chapter, we will first study the resonator equation without the transducer force in order gain a fundamental understanding of resonators. In Chapter 4, we will introduce mechanical nonlinearities to our system. Some elements of the analysis of linear and nonlinear resonators with nonlinear transducer forces will be introduced in Chapters 5 and 6.

Here, we present another interesting way to derive the formula for the transducer force, by employing the energy balance approach. The derivation of the force may seem a little complex in this case, however it does not depend on the configuration of the transducer. If required, you can apply this approach to any transducer – for example, electromagnetic, and derive the force that will appear in this case.

Assume that the electrostatic transducer is a device that has an "electrical" port and a "mechanical" port. The instantaneous electrical power flowing to the transducer at its electrical port is

$$P_e = V_t I_t \quad [3.12]$$

where V_t is the voltage across the transducer and I_t is the current flowing into the transducer. The instantaneous mechanical power at its mechanical port is

$$P_m = F_t \frac{dx}{dt} \quad [3.13]$$

where F_t is the mechanical force generated by the transducer and dx/dt is the velocity $v(t)$. Since we assume that there is no dissipation of energy in the transducer, electrical power P_e is spent on changing the energy of the transducer, W_t, and on generating mechanical power P_m at the output of the transducer:

$$P_e = \frac{dW_t}{dt} + P_m \qquad [3.14]$$

Using the relationship between transducer charge and voltage:

$$Q_t = C_t V_t \qquad [3.15]$$

where C_t is as usual the transducer capacitance, the current flowing through the transducer can be defined as:

$$I_t = \frac{dQ_t}{dt} = C_t \frac{dV_t}{dt} + V_t \frac{dC_t}{dx}\frac{dx}{dt} \qquad [3.16]$$

Combining [3.12] and [3.16], the instantaneous electrical power can now be written as:

$$P_e = C_t V_t \frac{dV_t}{dt} + V_t^2 \frac{dC_t}{dx}\frac{dx}{dt} = \frac{d}{dt}\left[\frac{C_t V_t^2}{2}\right] + \frac{1}{2} V_t^2 \frac{dC_t}{dx}\frac{dx}{dt} \qquad [3.17]$$

In this expression, we can recognize the electrical energy $W_e = C_t V_t^2/2$. Thus, the second term is mechanical power. Since we assume that the system is lossless and following on from [3.13] and [3.14], it is clear to see that

$$P_m(t) = \frac{V_t^2}{2}\frac{dC_t}{dx}\frac{dx}{dt} \qquad [3.18]$$

Recalling the definition of mechanical power through the mechanical force (power is the force multiplied by the velocity), we conclude that the mechanical force generated by the electromechanical transducer is given by formula [3.9].

Thus, the voltage generated by the conditioning electronics causes an additional force F_t on the mechanical resonator and the magnitude of this force is given by [3.9]. The general expression for F_t in form [3.9] will be the same for all electrostatic KEHs. However, each KEH will have its individual geometry of the capacitive transducer and its specific conditioning circuit. Therefore, the resulting expression for F_t for each individual realization of an electrostatic KEH will be different.

3.1.3. *Remarks on mechanical forces*

At this stage, we would like to make a remark on the nature of mechanical forces in the context of energy conversion.

In classical mechanics, a force describes an interaction between objects and it depends only on the position of the objects. Classical mechanics introduces a fundamental notion of *potential* (also called *conservative*) forces [LAN 76]. The work done by potential forces depends only on the starting and ending points x_1 and x_2 of the trajectory of a body and is independent of the shape of the trajectory. For such forces, we can introduce a quantity called potential energy that, in a one-dimensional case, is defined as the following integral:

$$W_p = -\int F(x)dx \qquad \text{or} \qquad F(x) = -\frac{dW_p}{dx} \tag{3.19}$$

where x is the position along the trajectory. It follows from this definition that the work done by a potential $F(x)$ on a body to move from x_1 to x_2 is $A_{x_1 \,\text{to}\, x_2} = \Phi(x_2) - \Phi(x_1)$ (where Φ is the antiderivative of F) and is independent of the trajectory.

Potential forces have a very important property: they only depend on the position x and not on the velocity of the body. A force generated by a linear spring can be seen as a good example of a potential force. This force depends only on the compression or expansion of the spring $F_{spring} = -k\Delta x$, where Δx is the deformation of the spring. In a more general case, a potential force may be a nonlinear function of the objects's position.

In an isolated mechanical system with only potential forces, total energy is conserved. For this reason, the potential forces are also called conservative.

However in practice, all mechanical systems experience a loss of energy, mainly due to different kinds of friction. For example, let us consider a body with an initial velocity v sliding on a rough uniform horizontal surface. It will eventually lose its momentum and stop. This behaviour is modeled by the introduction of a force that acts on the body and that depends only on its velocity. For example, the damper that we have introduced in the model of a linear resonator generates a linear friction force $F_{damping} = -\mu v$. In a more general case, a friction force can be a nonlinear function of the velocity.

Let us note that in the particular case of the air damping force we introduced, $\mu > 0$, and therefore energy decreases. In most real systems, damping forces reduce energy, and for this reason they are called *dissipative*. We must mention that there is the specific case of "negative" dissipation (in the case of a linear resonator, $\mu < 0$). It leads to the increase of the total energy of the system (such as in self-oscillators), however a discussion on self-oscillating systems is outside the scope of this book. Let us also note here that there is a collective term for such systems – *non-conservative*, since these forces modify the mechanical energy of the system.

Dissipative forces can also depend on both the displacement and the velocity. If the reader are familiar with the area of microelectromechanical systems (MEMS) and MEMS resonators, there is one known example of such forces – a squeeze-film damping force [SEN 01]. The transducer force, equation [3.9], is another example. From equation [3.9], we can see that it depends on the displacement through the term dC_t/dx. But it also depends on the velocity. Recall our discussion in Chapter 2 that energy is converted only if the capacitance decreases. The decrease of the capacitance also means the decrease of the relative displacement $x(t)$. Since the velocity is the derivative of $x(t)$, in the ideal conversion cycle we charge the transducer when $v(t) < 0$ and discharge when $v(t) > 0$. Thus, the dependance of F_t on the velocity is present.

We will give an illustration of conservative and non-conservative forces using the example of oscillatory applications. As we will learn later, KEHs typically display an oscillatory response and therefore the discussion below is directly related to KEH applications.

Consider a case when a body, whose position $x(t)$ is described by a sinusoidal function of time $x(t) = X_0 \sin(\omega t)$, is subjected to a nonlinear force $F(x(t), \dot{x}(t))$. In this case, the force $F(t)$ is also periodic, but, in the most general case, not sinusoidal. However, often only the first (fundamental) harmonic of the force is considered:

– in the case of a potential force $F = F(x(t))$ (dependent only on the dissipation), which can be nonlinear, the fundamental harmonic of the force has the same phase as the displacement (or with a π-phase shift), that is the force is also expressed as a sinusoidal function:

$$F(t) = F_1 \sin(\omega t), \tag{3.20}$$

We can show that F_1 is calculated as:

$$F_1 = \frac{1}{\pi}\left(\int_{-1}^{1} F(-X_0\sqrt{1-y^2})dy - \int_{-1}^{1} F(X_0\sqrt{1-y^2})dy\right) \tag{3.21}$$

Note that if $F(x)$ is even, the first harmonic of the force is equal to zero. However, this does not happen in most practical cases;

– in the case of a "purely" dissipative force $F = F(\dot{x}(t))$ (dependent only on the velocity), its fundamental harmonic has the same phase as the velocity (or with a π-phase shift), i.e., it is expressed as cosine function:

$$F(t) = F_1 \cos(\omega t), \quad [3.22]$$

where F_1 can be calculated as:

$$F_1 = \frac{1}{\pi}\left(\int_{-1}^{1} F(\omega X_0\sqrt{1-y^2})dy - \int_{-1}^{1} F(-\omega X_0\sqrt{1-y^2})dy\right) \quad [3.23]$$

Note that if $F(\dot{x})$ is even, the first harmonic of the force is equal to zero;

– in the case of the force that depends on both the displacement and velocity $F(x, \dot{x})$, the fundamental harmonic of the force may have a phase shift with respect to both the velocity and displacement:

$$F = F_1^c \cos(\omega t) + F_1^s \sin(\omega t) = F_1 \cos(\omega t + \phi_1) \quad [3.24]$$

The corresponding sine and cosine components are calculated as:

$$\begin{aligned} F_1^c &= \frac{1}{\pi}\left(\int_{-1}^{1} F(X_0 y, \omega X_0\sqrt{1-y^2})dy - \int_{-1}^{1} F(X_0 y, -\omega X_0\sqrt{1-y^2})dy\right) \\ F_1^s &= \frac{1}{\pi}\left(\int_{-1}^{1} F(X_0\sqrt{1-y^2}, \omega X_0 y)dy - \int_{-1}^{1} F(-X_0\sqrt{1-y^2}, \omega X_0 y)dy\right) \end{aligned} \quad [3.25]$$

The representation of the first harmonic of the force as the superposition of the sine and cosine components is sometimes interpreted as the superposition of potential and purely dissipative components (see the discussion in [SEN 01] on modeling of the squeeze-film damping force)[1]. It should be noted though that it is only useful in the context of systems with periodically varying displacements and velocities.

Thus, we want to highlight that the nature of the transducer force F_t is dissipative. It acts in a similar way as the damper force $F_{damping}$ and, stated simply, reduced the energy of the resonator. The spring force F_{spring}, which can also be nonlinear, is a potential force and does not lead to dissipation of energy.

1 The reader should be aware that if we consider the displacement in the form $x(t) = X_0 \cos(\omega t)$, the sine component of the force will represent, vice versa, the purely dissipative component while the cosine component will represent the potential component. In asymptotic and perturbation methods, this case of the displacement is more common. We will use this property in our discussion on the role of the transducer force in Chapter 6.

3.2. Interaction of the harvester with the environment

3.2.1. *Power balance of KEHs*

Having introduced all components of the energy harvester, we can now discuss the important question – the "power" balance of the energy harvester and the maximal power that can be converted by the harvester.

Let us consider the power balance of the energy harvester shown in Figure 3.2. We will describe the balance in terms of instantaneous power first. The power inflow is represented by the input power P_{in} that is brought to the system by external vibrations. This power inflow is equal to the change in the energy dW/dt stored in the resonator and the power outflow P_{out}. The power outflow consists of the power P_{damper} dissipated by the damper (the damper represents friction due to air or other sources of friction and dissipation) and the power P_t converted by the transducer force. Thus, the rate of change of the energy stored in the resonator is the sum of all power components:

$$\frac{dW}{dt} = P_{in} + P_{damper} + P_t \tag{3.26}$$

This equation does not take into account the nature the forces and the sign of their power. However, we assume that, on average, the external vibrations deliver the power to the resonator. In addition, according to our discussion presented in section 3.1.3, the transducer force and the damping force are non-conservative and, on average, removed the power from the resonator. (Thus, the average power due to $F_{damping}$ and F_t is lesser than zero, while the average power due to F_{ext} is positive).

We introduce the following averaging rule (with respect to time):

$$\overline{F(t)} = \frac{1}{T}\int_{t}^{t+T} F(\tau)d\tau \tag{3.27}$$

where T is the period of $F(t)$. In the case when $F(t)$ is not periodic, we should take the limit $T \to \infty$ in [3.27]. We derive the energy stored in a resonator and the energy relation of a linear resonator in later sections (sections 3.3.2 and 3.3.5) where we discuss the natural dynamics of the resonator. However, we note here that in a steady-state regime, when the resonator of a KEH displays forced oscillations, the average energy stored in the resonator does not change and so $\overline{dW/dt} = 0$.

Using this rule, we rewrite equation [3.26] in terms of average power:

$$\overline{P}_{in} + \overline{P}_{damper} + \overline{P}_t = 0 \tag{3.28}$$

From this expression, it is clear that the power $\overline{P}_t$ that can be converted by the harvester (transducer) is limited by P_{in}:

$$|\overline{P}_t| < \overline{P}_{in} \qquad [3.29]$$

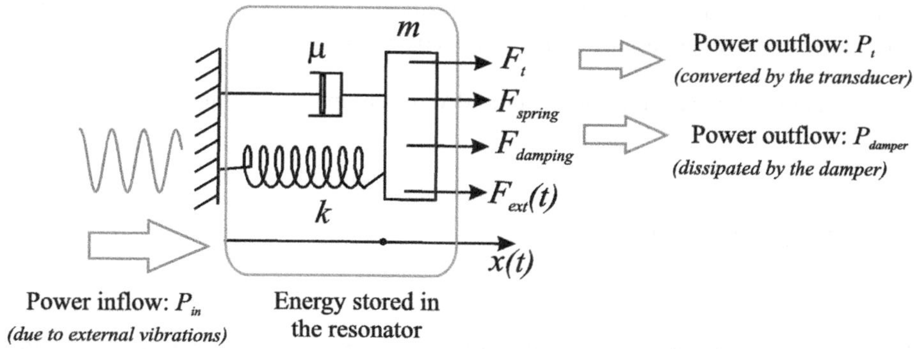

Figure 3.2. *Instantaneous power inflow and outflow in the energy harvester. The input power P_{in} that is brought to the system by external vibrations is equal to the change in the energy stored in the resonator dW/dt plus the power outflow P_{out}. The power outflow is represented by the power dissipated in the damper (friction due to air or other sources) and the power converted by the transducer*

Thus, it is useful to estimate the power that flows to the harvester from the environment since it will provide us with the *absolute limit* of the power we can convert with this harvester. We start by reminding the reader of the equation we obtained for the resonator of a KEH (equation [3.10]):

$$m\ddot{x} + b\dot{x} + kx = F_{ext} + F_t$$

Since we take periodic (moreover, harmonic) external vibrations

$$F_{ext} = \tilde{F}_0 \cos(\omega_{ext}t + \phi_0) = mA_{ext} \cos(\omega_{ext}t + \phi_0)$$

it is reasonable to assume that the response in the resonator will also be periodic (but not necessarily harmonic)[2]:

$$x(t) = \sum_{n=1}^{\infty} [a_n \cos(n\omega_{ext}t) + b_n \sin(n\omega_{ext}t)], \qquad \max[x(t)] \leq X_{lim} \qquad [3.30]$$

where X_{lim} is the limit on the range of the resonator displacement $x(t)$. This can be a physical limit (such as stoppers) or a design constraint.

By definition, the average power provided (or dissipated) by a mechanical force F is

$$\overline{P} = \frac{1}{T} \int_{t_0}^{t_0+T} F\, v\, dt \qquad [3.31]$$

where $v = dx/dt$ is the velocity. This is the average power due to the force F over the time T found from the instantaneous power $F \cdot v$ by averaging with respect to time. Note that in the case of aperiodic motion, one must again take the limit $T \to \infty$. In the case of periodic motion, T can be chosen simply as the period of $F(t)$. Taking into account that we want to find the average power $\overline{P}_{in}$ that is brought to the system by external vibrations, we assume $F = F_{ext}$. In addition, since the motion is periodic with the period $T = 2\pi/\omega_{ext}$, we write

$$\overline{P}_{in} = \frac{\omega_{ext}}{2\pi} \int_{0}^{2\pi/\omega_{ext}} F_{ext}\, \dot{x}\, dt \qquad [3.32]$$

For simplicity, we take $F_{ext} = mA_{ext} \cos(\omega_{ext}t)$ where we eliminated the initial phase shift ϕ_0 as discussed in the previous section. Substituting the velocity $v = dx/dt$ of the mass that is found by differentiating [3.30],

$$\frac{dx(t)}{dt} = \sum_{n} [-a_n\, n\omega_{ext} \sin(n\omega_{ext}t) + b_n\, n\omega_{ext} \cos(n\omega_{ext}t)] \qquad [3.33]$$

2 In the absence of nonlinear terms, such as F_t in equation [3.10], the response $x(t)$ in the resonator will be periodic and harmonic. However, in the presence of nonlinear terms, the response $x(t)$ becomes anharmonic, and we must take into account components other than the fundamental harmonic. We will discuss the linear and nonlinear responses in this and later chapters.

we obtain:

$$\overline{P}_{in} = \frac{mA_{ext}\omega_{ext}}{2\pi}$$

$$\times \int_0^{2\pi/\omega_{ext}} \cos(\omega_{ext}t) \sum_{n=1}^{\infty} [-a_n\, n\omega_{ext} \sin(n\omega_{ext}t) + b_n\, n\omega_{ext} \cos(n\omega_{ext}t)]\, dt$$

$$= \frac{mA_{ext}\omega_{ext}}{2\pi} \int_0^{2\pi/\omega_{ext}} \cos(\omega_{ext}t)[-a_1\omega_{ext} \sin(\omega_{ext}t) + b_1\omega_{ext} \cos(\omega_{ext}t)]dt$$

$$= \frac{mA_{ext}b_1\omega_{ext}}{2} \qquad [3.34]$$

Note that the cross-terms $\cos(\omega_{ext}t) \cdot \sin(n\omega_{ext}t)$ turn to zero for all n and $\cos(\omega_{ext}t) \cdot \cos(n\omega_{ext}t)$ turn to zero for $n \neq 1$ as a result of the integration.

Finally, we recall that we want to estimate the maximum possible power. A periodic function is bounded by the limits X_{lim}: $-X_{lim} \leq x \leq X_{lim}$. In order to estimate maximal possible b_1, we assume that the displacement has a square waveform ($n = 0, 2, 4 \ldots$):

$$x(t) = \begin{cases} X_{lim}, & n\frac{T}{2} \leq t < (n+1)\frac{T}{2} \\ -X_{lim}, & (n+1)\frac{T}{2} \leq t < (n+2)\frac{T}{2} \end{cases} \qquad [3.35]$$

as it will provide us with the largest possible value of b_1. We compute it as follows:

$$\begin{aligned} b_1 &= \tfrac{2}{T} \int_0^T x(t) \sin(2\pi/T)dt \\ &= \tfrac{2}{T} \int_0^{T/2} X_{lim} \sin(2\pi/T)dt + \tfrac{2}{T} \int_{T/2}^{T} (-X_{lim}) \sin(2\pi/T)dt \\ &= \tfrac{4X_{lim}}{\pi} \end{aligned} \qquad [3.36]$$

Thus, with $b_{1,max} = 4X_{lim}/\pi$, we obtain the final formula:

$$\overline{P}_{in,max} = \frac{2mA_{ext}X_{lim}\omega_{ext}}{\pi} \qquad [3.37]$$

This formula gives the absolute limit on the power that can be converted from sinusoidal external mechanical vibrations provided that the displacement of the resonator is periodic. It clearly shows that in order to maximise the power, we should increase the amplitude of external vibrations A_{ext}, the frequency ω_{ext} or/and the range of the resonator displacement X_{lim}.

3.2.2. *Efficiency of KEHs*

Having derived the input power [3.37], we are going to show how this can be used to correctly define *KEH efficiency*.

First, we note that there is no universal definition of efficiency. In a majority of power conversion systems, efficiency η is defined as the ratio of the power taken at the output and the power supplied at the input of the system:

$$\eta = \frac{P_{out}}{P_{in}} \qquad [3.38]$$

We are going to show that such a definition is not convenient (and, moreover, misleading) for energy harvesting systems.

Let us consider a simple example – a realistic model of a voltage source that consists of a source V_S in series with an internal resistance R_S connected to a resistive load R_L (as shown in Figure 3.3). We know that in order to transfer maximal power from V_S to R_L, R_L must be equal to R_S.

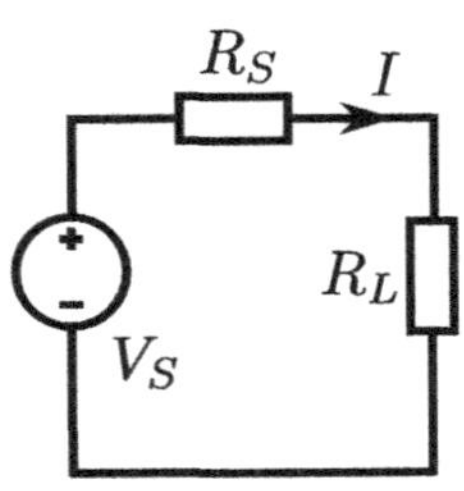

Figure 3.3. *Elementary circuit illustrating the transmission of power from a voltage source V_S to a load resistance R_L*

However, in a limiting case when the system is clearly inefficient, that is $R_L \gg R_S$, from [3.38] we obtain that $\eta \approx 1$:

$$\frac{P_{R_L}}{P_{V_S}} = \frac{I^2 \cdot R_L}{V_S \cdot I} = \frac{R_L}{R_L + R_S} \to 1 \quad \text{if} \quad R_L \gg R_S, \qquad [3.39]$$

In this expression, I is the current flowing in the circuit through both resistors. Although efficiency defined as [3.38] approaches the ideal case, the system is useless. Only a very small portion of power can be realistically transmitted from V_S to R_L in this case.

What is wrong with the definition [3.38] when applied to this case? This definition does not account for the limited capability of the system to extract energy from the voltage source. Indeed, although an ideal voltage source is, theoretically, an infinite source of energy, when connected through an internal resistance it can provide only a limited amount of energy.

A useful definition of efficiency in our case will be the ratio of the power delivered to the load and the maximal power that can be delivered when the load is matched with the internal resistance of the source:

$$\eta = \frac{P_{out}}{P_{in,max}} \qquad [3.40]$$

Adopting this reasoning to the case of energy harvesting systems, we can state that miniature (micro-scale) KEHs have almost infinite amount of energy at their input. If one considers the vibration of a car or even the motion of a human body, the available mechanical energy is huge compared to the amount that can be harvested by a miniature harvester. This can also be understood from the following consideration: since energy harvesting devices should not impede or affect the motion of the main system (e.g. a car), the harvested power must be negligible compared to the power generated from the "normal" operation of the ambient system. As a consequence, the original efficiency definition [3.38] is useless, since it will give zero efficiency in all cases.

But what is the maximal input power that can ideally be extracted from the mechanical domain by a harvester? We have already obtained this value in the previous section, and it is given by equation [3.37]. This maximum is independent of the technology used for electromechanical conversion. Thus,

$$\eta_{KEH} = \frac{\text{Average electrical power delivered to the load}}{\overline{P}_{in,max}} \qquad [3.41]$$

The maximal power [3.37] is defined by two parameters describing the harvester (m and X_{lim}) and two parameters dictated by the vibration source (ω_{ext} and A_{ext}). We can also note that the mass m is proportional to the volume, and so the harvested power is proportional to the characteristic size of the device as follows: $P \sim X^4$. The size of the KEH plays an important role: one should not anticipate a large amount of power at the output of a miniature KEH unless it oscillate at a very large frequency ω_{ext}.

Now, we will proceed to the study of the natural dynamics of the resonator as it provides us with some idea on how $x(t)$ behaves with time and what X_{lim} we can obtain.

3.3. Natural dynamics of the linear resonator

3.3.1. *Behavior of the resonator with no input*

We start the analysis of the linear resonator with a brief discussion on the behaviour of the resonator if no input is applied. Indeed, with no external vibrations, what would be the output of a KEH? In this section, we are aiming to show that external vibrations are required to get some useful output form a KEH – without them, the KEH resonator does not display any motion. All the expressions we will provide in this section are required for the fundamental understanding of the behavior of resonators.

We consider a damped resonator given by [3.6] with $\gamma > 0$ with no input, that is $F_{ext} = 0$. We solve equation [3.6] using the standard method, that is by using the Euler substitution $x = \exp(\lambda t)$, where λ is the characteristic number, and taking the right-hand side to be zero [KRE 88]. Here, we give the final result yielded by this method.

For sufficiently small dissipation $\gamma < \omega_0$ the solution is:

$$\begin{aligned} x(t) &= e^{-\gamma t}\left[x_0 \cos \omega t + \frac{v_0 + \gamma x_0}{\omega} \sin \omega t\right] \\ v(t) &= e^{-\gamma t}\left[v_0 \cos \omega t - \frac{\gamma v_0 + \omega_0^2 x_0}{\omega} \sin \omega t\right] \end{aligned} \qquad [3.42]$$

where we have introduced a new quantity $\omega = \sqrt{\omega_0^2 - \gamma^2}$ (note that $\omega_0^2 - \gamma^2 > 0$, and therefore ω is a real value). It is sometimes referred to as the frequency of the damped oscillatory process. In the equation above, we used the initial conditions $x(0) = x_0$ and $\dot{x}(0) = v_0$.

For sufficiently large dissipation $\gamma > \omega_0$ on the other hand, the solution is:

$$\begin{aligned} x(t) &= e^{-\gamma t}\left[x_0 \cosh \varpi t + \frac{v_0 + \gamma x_0}{\varpi} \sinh \varpi t\right] \\ v(t) &= e^{-\gamma t}\left[v_0 \cosh \varpi t - \frac{\gamma v_0 + \omega_0^2 x_0}{\varpi} \sinh \varpi t\right] \end{aligned} \qquad [3.43]$$

where we have introduced another quantity $\varpi = \sqrt{\gamma^2 - \omega_0^2}$ (in this case, $\gamma^2 - \omega_0^2 > 0$ and therefore ϖ is a real value). In contrast to equations [3.42] that describe decaying oscillating process, equations [3.43] describe a straightforward non-oscillatory decay of the displacement and velocity with time since these equations contain hyperbolic sine and cosine functions ($\sinh(\varpi t)$ and $\cosh(\varpi t)$). The displacement, $x(t)$, of a damped linear resonator is shown in Figure 3.4(a), where the three curves show a linear resonators with different dissipation (underdamped and overdamped). Resonators with $\omega_0 > \gamma$ or $Q > 1/2$ are called underdamped while resonators with $\omega_0 < \gamma$ or $Q < 1/2$ are called overdamped.

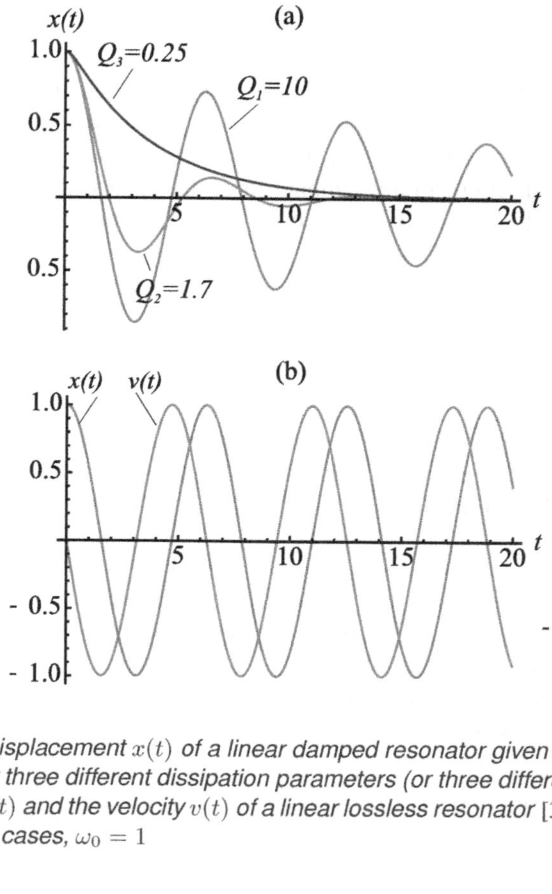

Figure 3.4. *a) Displacement $x(t)$ of a linear damped resonator given by equation [3.6] with $F_{ext} = 0$ for three different dissipation parameters (or three different Q-factors); b) Displacement $x(t)$ and the velocity $v(t)$ of a linear lossless resonator [3.6] with $F_{ext} = 0$ and $\gamma = 0$. In all cases, $\omega_0 = 1$*

We must mention here that in the special case of the conservative resonator when $\gamma = 0$, the solution of differential equation [3.6] is:

$$\begin{aligned} x(t) &= x_0 \cos \omega_0 t + \tfrac{v_0}{\omega_0} \sin \omega_0 t = A_0 \cos(\omega_0 t + \varphi_0) \\ v(t) &= v_0 \cos \omega_0 t - x_0 \omega_0 \sin \omega_0 t = -A_0 \omega_0 \sin(\omega_0 t + \varphi_0) \end{aligned} \quad [3.44]$$

The quantity $A_0 = \sqrt{x_0^2 + \frac{v_0^2}{\omega_0^2}}$ defines the amplitude of oscillations, while the quantity $\tan \varphi_0 = (-v_0/\omega_0 x_0)$ defines their phase. This solution is a non-decaying harmonic oscillation as shown in Figure 3.4.

Note, however, that it is not possible to achieve zero dissipation in realistic systems. Therefore, the natural behavior of the realistic resonator is a decaying oscillatory process that leads to $x(t) = v(t) = 0$ after sufficiently long time t. If the initial conditions of the resonator are zero, there will be no motion.

3.3.2. *Energy relation for the resonator with no input*

Let us obtain the energy relation of the damped linear resonator. Using this equation, we want to show that even if there were some initial energy stored in the resonator (e.g. due to the presence of the initial velocity or displacement), it would have dissipated from it due to the damping force that represents friction and appears in the equation in the form of the the term $\mu \dot{x}$ (or $\gamma \dot{x}$). Let us multiply equation [3.6] (with F_{ext} set to zero) by $\dot{x}$. We obtain

$$\frac{d}{dt}\left[\frac{m\dot{x}^2}{2} + \frac{kx^2}{2}\right] = -\mu\dot{x}^2 \qquad [3.45]$$

The term $m\dot{x}^2/2$ represents the instantaneous kinetic energy W_k of the proof mass while the term $kx^2/2$ represents the instantaneous potential energy W_p of the spring (see equation [3.19]). Together they constitute the total energy of the resonator $W = W_k + W_p$.

From this equation, we can see that the energy does not remain constant and decreases due to the action of dissipation. We can rewrite the above expression in another form. By definition, power is $P = dA/dt = vF$, where v is the velocity of the point where the force is applied, F is some force and A is the work done by the force F. We can write that

$$\frac{dW}{dt} = \frac{d}{dt}\left[W_k + W_p\right] = P_{damper} \qquad [3.46]$$

where $P_{damper} = -\mu\dot{x}^2 = -\mu v^2$ is the power of the friction force. Note that $F_{damping}$ is a dissipative force and so $P_{damper} < 0$ leading to the decrease of resonator energy.

Finally, we can take into account the action of the transducer force, recalling its dissipative nature. Taking equation [3.11] and repeating the steps described above, we obtain:

$$\frac{dW}{dt} = \frac{d}{dt}\left[W_k + W_p\right] = P_{damper} + P_t \qquad [3.47]$$

With the transducer connected to the resonator, its energy dissipates even faster.

It is clear that without the inflow of power (in our case – due to external mechanical vibrations represented by the force F_{ext}), all the energy in the resonator will dissipate. In order to sustain the motion of the resonator and convert energy into the electrical domain, we must provide this external force.

3.3.3. *Forced oscillator and linear resonance*

Therefore, as we established, we must apply an external force in order to obtain motion in the linear resonator. In the context of KEHs, the external force will be provided by the environment in the form of periodic or non-periodic vibrations. As we agreed in this chapter, we will study in detail the case of the external force given by equation [3.3]. Let us recall that the relative displacement of the resonator (its proof mass) is described as

$$\ddot{x} + 2\gamma\dot{x} + \omega_0^2 x = F_0 \cos(\omega_{ext} t + \phi_0) \qquad [3.48]$$

where $F_0 = \tilde{F}_0/m = A_{ext}$ is the amplitude of the external vibrations (as follows from [3.3]).

The solution of this differential equation is a superposition of a solution corresponding to the homogeneous equation [3.6] and of a particular solution of [3.48] (see, e.g. [KRE 88]):

$$x(t) = x_h(t) + x_{nh}(t) \qquad [3.49]$$

where $x_h(t)$ is known from [3.42] or from [3.43].

The force on the right-hand side of equation [3.48] can be presented in the form $F(t) = \text{Re}[F_0 \exp(j\phi_0) \exp(j\omega_{ext} t)]$ and the quantity $\mathbb{F} = F_0 \exp(j\phi_0)$ is called the complex amplitude of the force or the force phasor. Because of the specific form of this force, we will be looking for a particular solution of the non-homogeneous equation [3.48] in the form $x(t) = \text{Re}[A \exp(j\phi) \exp(j\omega_{ext} t)]$ where A and ϕ are the resulting amplitude and phase of forced oscillations. Here, j is the imaginary unit, and it satisfies the equation $j^2 = -1$. Correspondingly, the quantity $\mathbb{X} = A\exp(j\phi)$ is called the complex amplitude or the phasor of the displacement (or more generally, of a generic signal).

After substituting x and F into the original equation, we obtain the equation relating the two complex amplitudes:

$$(-\omega_{ext}^2 + 2j\gamma\omega_{ext} + \omega_0^2)\mathbb{X} = \mathbb{F} \tag{3.50}$$

We will skip the further steps of the derivation as they are straightforward. The resulting solution is:

$$x_{nh}(t) = A\cos(\omega_{ext}t + \phi + \phi_0) \tag{3.51}$$

In this expression, the amplitude A of the forced oscillation is defined as:

$$A = \frac{F_0}{\sqrt{(\omega_0^2 - \omega_{ext}^2)^2 + 4\gamma^2\omega_{ext}^2}} \tag{3.52}$$

while the the phase ϕ of the forced oscillation is found from the following expressions:

$$\cos\phi = \frac{\omega_0^2 - \omega_{ext}^2}{\sqrt{(\omega_0^2 - \omega_{ext}^2)^2 + 4\gamma^2\omega_{ext}^2}}, \qquad \sin\phi = \frac{-2\gamma\omega_{ext}}{\sqrt{(\omega_0^2 - \omega_{ext}^2)^2 + 4\gamma^2\omega_{ext}^2}} \tag{3.53}$$

Now we write the full solution of this system that represents the superposition of oscillations [3.42] and forced oscillations [3.51]:

$$x(t) = x_h(t) + x_{nh}(t) = e^{-\gamma t}\left[C_1\cos\omega t + C_2\sin\omega t\right] + \frac{F_0}{\sqrt{(\omega_0^2 - \omega_{ext}^2)^2 + 4\gamma^2\omega_{ext}^2}}\cos(\omega_{ext}t + \phi + \phi_0) \tag{3.54}$$

where the coefficients C_1 and C_2 are defined from the initial conditions:

$$C_1 = x_0 - x_{nh}(0), \quad C_2 = \frac{v_0 - \dot{x}_{nh}(0) + \gamma(x_0 - x_{nh}(0))}{\omega} \tag{3.55}$$

and x_0 and v_0 denote, as usual, the initial state of the resonator at $t = 0$.

It is important to mention that after a sufficiently long time, the resonator will display only the forced oscillations. As follows from [3.54], the eigen oscillation term is multiplied by $\exp(-\gamma t)$, and if time t is large, this term tends to zero. Thus, we may assume that [3.51] describes the steady-state regime of a forced linear resonator.

Let us investigate the amplitude of forced oscillations A given by [3.52]. It depends on the parameters of the external force (the force amplitude F_0 and the frequency ω_{ext}) and on the parameters of the resonator itself (its natural frequency ω_0 and the dissipation factor γ).

It is obvious that for a particular resonator with fixed ω_0 and γ, the dependence on the force amplitude F_0 is linear: as F_0 increases, the larger the response of the resonator becomes. However, the dependence of A on the external frequency ω_{ext} is somewhat non-trivial (see Figure 3.5). The amplitude A as a function of ω_{ext} displays an increase in the resonator response to the external force in the case when the frequency ω_{ext} of the external force approached the natural frequency of the resonator, that is $\omega_{ext} \approx \omega_0$. This increase is more visible for resonators with higher Q-factors and less visible for resonators with lower Q-factor.

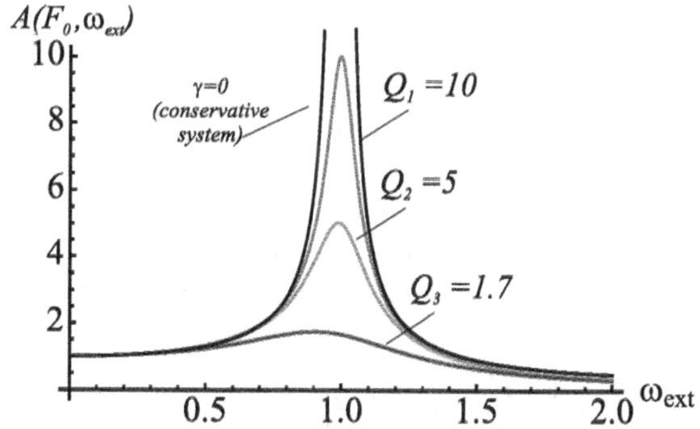

Figure 3.5. *The amplitude of the forced oscillations of a linear damped resonator $A(F_0, \omega_{ext})$ as a function of the external frequency ω_{ext} for three different Q-factors. In the case of a conservative linear resonator with $\gamma = 0$, the amplitude of forced oscillations increases infinitely: $\omega_0 = 1$, $F_0 = 1$*

This phenomenon is called *resonance*. On a qualitative level, it can be defined as an increase in the amplitude of forced oscillations when the frequency of the external force ω_{ext} is close to the natural frequency of the resonator ω_0. For a conservative resonator with $\gamma = 0$, the amplitude of forced oscillations will increase infinitely, as shown in Figure 3.5.

It is interesting to note that at $\omega_{ext} = \omega_0$, the phase shift of forced oscillation found from [3.53] is exactly $-\pi/2$:

$$\cos\phi = 0, \quad \sin\phi = -1 \quad \Longrightarrow \quad \phi = -\pi/2 \qquad [3.56]$$

Sometimes, the case $\omega_{ext} = \omega_0$ is called *exact resonance* in the literature. Note that the maximum of the resonator displacement (i.e. the maximum of A in [3.52]) is reached when $\omega_{ext} = \omega_0\sqrt{1 - 2\gamma^2/\omega_0^2}$. You can obtain this by treating A as a function of ω_{ext} and finding the extremum of this function (i.e. setting the derivative $dA/d\omega_{ext}$ to zero).

On the other hand, the maximum of the velocity $v = \dot{x}$ occurs at a different frequency. This is because the displacement and velocity are related as $v = -\omega_{ext}\, x$ and so the "amplitude" of the velocity waveform in the resonator is

$$A_v = \frac{-\omega_{ext} F_0}{\sqrt{(\omega_0^2 - \omega_{ext}^2)^2 + 4\gamma^2\omega_{ext}^2}} \qquad [3.57]$$

The maximum of this A_v is reached at $\omega_{ext} = \omega_0$, that is at exact resonance.

The amplitude $A(F_0, \omega_{ext})$ is directly related to the transfer function of a resonator. Such an approach, utilizing transfer functions, may be more familiar to readers with a background in electric and electronic engineering. In order to obtain the transfer function of a resonator, we use the substitution $x = X_0 \exp(st)$ into equation [3.48] (we use that $F = F_0 \exp(st)$), as a result we obtain:

$$X_0 = H(s)F_0 = \frac{F_0}{s^2 + 2\gamma s + \omega_0^2} \qquad [3.58]$$

The resonator can be seen as a filter (see Figure 3.6) with the transfer function shown in Figure 3.5. In particular, an underdamped resonator with $Q > 1/2$ responds effectively only to harmonic oscillations whose frequency $\omega_{ext} \approx \omega_0$. Moreover, the higher the Q-factor of the resonator is, the more effective the response at the fundamental frequency ω_0 is. In a broader sense, if the input is a periodic force with a number of harmonic components in it, the resonator will select only those harmonics whose frequencies are close to its natural frequency ω_0.

The bandwidth of the resonator is often defined as the width of the resonance curve in Figure 3.5 taken at $A = A_{max}/\sqrt{2}$:

$$\text{BW} = \Delta\omega = \frac{\omega_0}{Q} = 2\gamma \qquad [3.59]$$

where we remind you that Q is the quality factor. (However, other definitions of the bandwidth also exist.)

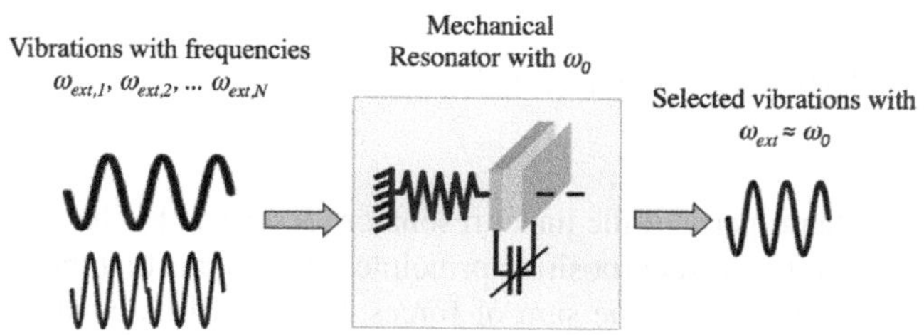

Figure 3.6. *The linear resonator acts as a filter. If a superposition of harmonic vibrations with different frequencies $\omega_{ext,i}$ is applied at its input, it will select a vibration (or vibrations) with only a frequency (frequencies) $\omega_{ext} \approx \omega_0$*

It is obvious that for large Q, the resonator becomes highly selective. Its bandwidth BW decreases and the resonator filters out all frequencies except $\omega_{ext} \simeq \omega_0$. In addition, from [3.52], it follows that for larger Q, the amplitude of the resonator response significantly increases.

Therefore, using the approximation of a high-Q resonator ($Q \gg 1$), we can simplify and write the response of a resonator to harmonic oscillations of the form $mA_{ext} \cos(A_{ext}\omega_{ext}t)$:

$$x(t) \approx \frac{A_{ext}\, Q}{\omega_0^2} \sin(\omega_{ext}t) \qquad [3.60]$$

where we used that $\phi_0 = 0$ (recall that it is always possible to eliminate it), $\omega_{ext} \approx \omega_0$ (otherwise there will be no response in a high-Q resonator) with $\phi \approx -\pi/2$. This simplified expression is very often used in analytical analysis of KEHs.

3.3.4. *Periodic external vibrations*

We can also take into account a more general form of external vibrations. For example, multi-modal external vibrations or some periodic (but not necessarily harmonic) force. In both cases, we will take F_{ext} in the form of equation [3.4]. Thus, we are to solve the equation:

$$m\ddot{x} + \mu\dot{x} + kx = \sum_{n=1}^{N} mA_{ext,n}\cos(\omega_{ext,n}t + \phi_{0,n}) \qquad [3.61]$$

One remarkable property of the linear resonator (described by the above equation) is that it is subjected to the superposition principle. This means that the total response of the linear resonator $x(t)$ to the sum of forces $F_{ext,n}$ will be equal to the sum of individual responses x_n to each force $F_{ext,n}$.

We have already found the steady-state response of the linear resonator to a single external harmonic oscillation – it is given by [3.51], [3.52] and [3.53]. Thus,

$$x(t) = \sum_{n=1}^{N} x_n = \sum_{n=1}^{N} A_n\cos(\omega_{ext,n}t + \phi_n + \phi_{0,n}) \qquad [3.62]$$

where A_n and ϕ_n are found from [3.52] and [3.53] by taking appropriate $A_{ext,n}$, $\omega_{ext,n}$ and $\phi_{0,n}$ in the relevant expressions.

3.3.5. *Energy relation for a forced resonator*

We will conclude the study of linear resonators with the energy relation for a forced harmonic resonator, showing that the external vibration will deliver energy to the resonator.

We will take equation [3.48] and multiply it by $v = \dot{x}$:

$$\frac{d}{dt}\left[\frac{mv^2}{2} + \frac{kx^2}{2}\right] = vF_{ext}(t) - \mu v^2 \qquad [3.63]$$

where $vF_{ext}(t)$ expresses the instantaneous power of the external force and $-\mu v^2$ expresses the instantaneous power dissipated by the damper. The above equation describes the energy conservation law in this system: the power of the external force is dissipated by the change of the energy of the resonator and losses due to dissipation.

We will find the average of [3.63] over the period $T = 2\pi/\omega_{ext}$ by applying the averaging rule [3.27]. With regard to the term

$$\frac{d}{dt}\left[\frac{mv^2}{2} + \frac{kx^2}{2}\right]$$

we recall that the displacement x and the velocity v are harmonic oscillations. We can take $x(t) = A\cos(\omega_{ext}\, t + \phi)$ and $v(t) = -\omega_{ext}\, A\sin(\omega_{ext}\, t + \phi)$. Since

$$\overline{W_k} = \overline{mv^2/2} = \frac{m\omega_{ext}}{4\pi}\int_0^{2\pi/\omega_{ext}} \omega_{ext}^2 A^2 \sin^2(\omega_{ext}\tau + \phi)d\tau = m\omega_{ext}^2 A^2/4 \qquad [3.64]$$

and

$$\overline{W_p} = \overline{kx^2/2} = \frac{m\omega_{ext}}{4\pi}\int_0^{2\pi/\omega_{ext}} A^2 \cos^2(\omega_{ext}\tau + \phi)d\tau = kA^2/4 \qquad [3.65]$$

We see that the total average energy $\overline{W} = \overline{W_k} + \overline{W_p}$ is constant, and so its derivative with respect to time is zero: $\overline{dW/dt} = 0$.

Thus, we obtain

$$\overline{P}_{damper} + \overline{P}_{in} = 0 \qquad [3.66]$$

where $\overline{P}_{damper} = \overline{-\mu v^2(t)}$ and $\overline{P}_{in} = \overline{v(t)F_{ext}(t)}$. We can add the transducer force in the same manner as we did in section 3.3.2 and write

$$\overline{P}_{damper} + \overline{P}_t + \overline{P}_{in} = 0 \qquad [3.67]$$

where $\overline{P}_t = \overline{v(t)\ F_t(t)}$. As a matter of fact, we obtained the same result as in [3.28].

Equations [3.66] and [3.67] express the fact that the average power of the external force, over the period of oscillations, is equal to the power dissipated by the friction forces acting on the system (including the dissipative transducer force F_t). Thus, the energy of the forced resonator does not dissipate to zero with time: the external force delivers the energy from the mechanical domain to compensate the losses.

The phenomenon of forced oscillations and resonance lies at the heart of vibration energy harvesting. External vibrations serve as the external force that drives the motion of the KEH resonator. In order to obtain a sufficient, large response, that is large amplitude vibrations in the resonators, we must ensure that (i) the resonator has a high-quality factor and (ii) the frequency (or frequencies) of external vibrations is close to the natural frequency of the resonator to achieve resonance. The external vibrations cause the displacement of the resonator and therefore of the transducer, thus giving rise to the energy conversion cycle that converts energy from the mechanical to electrical domain.

3.4. The mechanical impedance

At this point, we will introduce a useful tool for the representation and analysis of mechanical systems, which is based on the notion of the mechanical impedance. This approach is inspired by the well-known mathematical analogy between electrical networks and mechanical systems with lumped parameters. An example of equivalence between mechanical and electrical quantities is given in Table 3.1. This equivalence is derived from the fact that both a mass-spring-damper system and an RLC electrical resonator are governed by a second-order linear differential equation of form [3.2].

Mechanical quantities	Electrical quantities
Force (F)	Electromotive force ($\mathcal{E}$), also denoted a voltage ($-V$ or $-u$)
Velocity (v)	Current (i)
Position (x)	Charge (q)
Mass (m)	Inductance (L)
Stiffness (k)	Electrical elastance ($1/C$)
Damping coefficient (μ)	Resistance (R)

Table 3.1. *Equivalence between mechanical and electrical quantities*

The force F_{ext} is represented by an independent voltage source, while the velocity of the proof mass by an electrical current. The transducer is represented by an electrical dipole whose voltage (the inertial force $-m\ddot{x}$) depends on the velocity of the proof mass (i.e. the current). The mechanical resonator is then equivalent to a series RLC network, and a KEH can be represented by the electrical circuit of Figure 3.7.

Figure 3.7 describes the system in the mechanical domain. However, the transducer force F_t depends also on the architecture of the conditioning circuits (i.e. on its voltage, charges, electrical or magnetic fields, and etc.). Thus, as discussed in Chapter 1, a KEH must be understood as a coupled system that consists of the resonator, the transducer and the conditioning circuit, and not as an isolated device interfacing electrical and mechanical sub-systems.

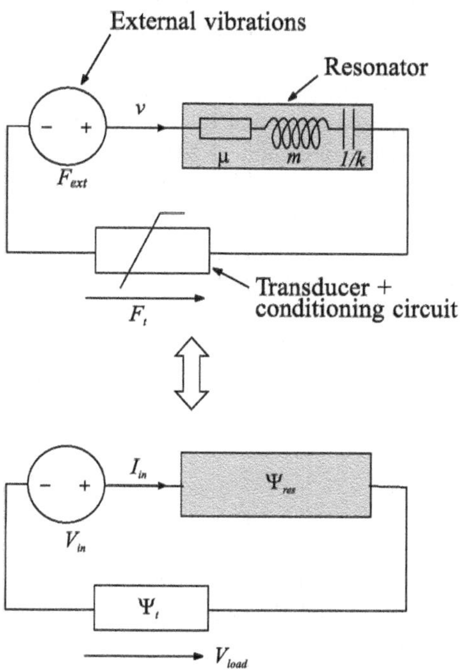

Figure 3.7. *Equivalent electrical representation of a KEH*

In linear electrical systems, the impedance of a dipole is defined as the ratio between the phasor of the voltage taken across its terminals and the phasor of the current flowing through it. The impedance Ψ of a mechanical system is defined equivalently to its electrical counterpart:

$$\Psi = -\frac{\mathbb{F}}{\mathbb{V}} \quad [3.68]$$

where $\mathbb{F}$ is the phasor of a harmonic mechanical force and $\mathbb{V}$ is the phasor of the velocity. Following this analogy, the mechanical impedance of a KEH resonator is given by:

$$\Psi_r = \frac{k}{j\omega} + j\omega m + \mu \quad [3.69]$$

Similarly, the ratio between the phasor of the transducer force $\mathbb{F}_t$ and the velocity of the proof mass $\mathbb{V}$ defines the mechanical impedance Ψ_t of the electromechanical transducer plus its conditioning circuit. The VEH can now be represented by only two mechanical impedances, Ψ_r and Ψ_t, as shown in Figure 3.7.

In the literature on kinetic energy harvesting, the electromechanical transducer is often represented by an ideal damper, that is as a dipole that has a real impedance. While this is the case for electromagnetic KEHs, it cannot be generalized to other transduction mechanisms. In general, the impedance of the transducer is complex and need to be optimized so that the power dissipated in the load is maximum.

In electrical networks, the power dissipated in a dipole is the real part of the complex power, which is defined as the half-square of the modulus of the current times the impedance of the load. Since in the mechanical domain the current is equivalent to the velocity, and the mechanical impedance is defined by equation [3.68], the power converted by the transducer is

$$P_t = \frac{1}{2}\mathbb{V}^2\mathrm{Re}(\Psi_t) \qquad [3.70]$$

The mechanical impedance methods is a convenient design tool that has been applied to the analysis of eKEHs with nonlinear forces F_{ext} in studies [GAL 11a, GAL 13a, BAS 14].

3.5. Concluding remarks

Therefore, in this chapter, we have derived the model equation of a linear resonator. We have discussed the nature of forces and their effects on the resonator and the energy stored in it. The model of the linear resonator allows us to introduce and quantify linear resonance. The phenomenon of forced oscillations and resonance provides the basis for kinetic energy harvesting applications: "amplification" oscillations in a resonator are used to convert energy from the mechanical to electrical domain. In order to exploit resonance, we must ensure that the resonator has a high quality factor. This, however, will limit us to a narrow range of frequencies of external vibrations $\omega_{ext} \approx \omega_0$ since a high-Q resonator acts as a selective filter, effectively reducing the response to all other frequencies. In order to deal with this limitation, researchers introduce nonlinearities, mechanical or electrical, that allow us to expand the frequency response of resonators. We discuss nonlinear resonators in the next chapter.

4

Mechanical Aspects of Kinetic Energy Harvesters: Nonlinear Resonators

The aim of this chapter is to explain the concept of a nonlinear resonator. We discuss resonance in nonlinear resonators, multistability of the potential function and multistability of the nonlinear resonance curve (which we will call multimodality throughout the book). We explain how these are used in energy harvesting applications.

4.1. Nonlinear resonators with mechanically induced nonlinearities

4.1.1. *Equation of the nonlinear resonator*

Implicitly, we make a number of assumptions when we derive the equation of the linear resonator [3.6]. In particular, one of our key assumption concerns the spring restoring force. We say that the spring that models the elastic restoring properties of the resonator is characterized by a constant coefficient k so that the restoring force is linear:

$$F_{spring} = -kx \qquad [4.1]$$

Such a spring is also called linear. It is a good assumption that follows from Hooke's law. This works very well for small displacements, but for larger displacements we start seeing some discrepancies.

Suppose that the actual restoring force is described as an unknown nonlinear function of the displacement x:

$$F_{spring} = \mathcal{F}(x) \qquad [4.2]$$

Therefore, we will rewrite equation [3.6]:

$$\ddot{x} + 2\gamma\dot{x} + g(x) = F_{ext}/m \tag{4.3}$$

where $g(x) = -\mathcal{F}(x)/m$.

We can make one reasonable assumption about the spring force. Since the spring force is usually restoring, it is directed in such a way that it compensates an increase in the displacement x. Indeed, the linear spring force $F_{spring} = -kx$ is positive for $x < 0$ and negative for $x > 0$. Thus, F_{spring} is an odd function of x. Generalizing to the case of the nonlinear spring function, we assume that $\mathcal{F}(-x) = -\mathcal{F}(x)$. We can write the Taylor series for equation [4.2] around the zero displacement:

$$F_{spring} \approx \frac{\mathcal{F}'(0)}{1!}x + \frac{\mathcal{F}'''(0)}{3!}x^3 + \frac{\mathcal{F}^{(v)}(0)}{5!}x^5 + ... \approx -\left[k_1x + k_3x^3 + ...\right] \tag{4.4}$$

where $k_1 = -\mathcal{F}'(0)$, $k_3 = -\mathcal{F}'''(0)/(3!)$, etc. This series will have only odd orders of x since $\mathcal{F}(x)$ is odd. The minus sign in equation [4.4] explicitly highlights that the force $\mathcal{F}(x)$ is restoring (for the same reason, we write $F_{spring} = -kx$). Resonators with this type of nonlinearity (i.e. expressed by odd functions) constitute a major and very important class of nonlinear resonators and are very typical for mechanical systems.

Note, by the way, that the coefficients k_3, k_5 and others can be of any sign (positive or negative). However, we assume that $k_1 > 0$. Indeed, our series [4.4] should return the linear spring described by Hook's law for small x. This is always the case for elastic structures such as springs and arms that have elastic properties. The case of $k_1 < 0$ is very specific: such systems have or are artificially designed to have an unstable equilibrium at $x = 0$. We will discuss them later in this chapter.

In particular, if the series [4.4] is truncated at the term with x^3, we obtain a famous model with cubic nonlinearity. It is called the *Duffing resonator* [DUF 18]. Historically, it was one of the first models of nonlinear resonators:

$$\ddot{x} + 2\gamma\dot{x} + \omega_0^2x + \kappa x^3 = F_{ext}/m \tag{4.5}$$

where $\omega_0^2 = k_1/m$ and $\kappa = k_3/m$ (can be of any sign). Note that it is often accepted, following the linear resonator model, that the coefficient appearing before the linear term x is denoted as ω_0^2. The Duffing type of nonlinearity appears very often in microscale mechanical structures (MEMS) induced by their mechanical properties [ADA 98, DEM 07, SEN 01]. Clearly, mechanical nonlinearity is more noticeable if the displacement x is large.

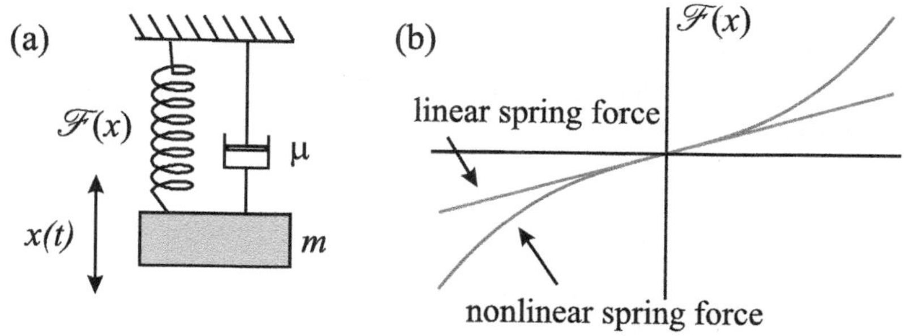

Figure 4.1. *a) Simple nonlinear mechanical resonator that is obtained from the linear resonator in Figure 3.1 by considering that the spring restoring force $F_{spring}(x)$ has a nonlinear dependence on the displacement x; b) Example of nonlinear spring force in the form $F_{spring}(x) = k_1x + k_3x^3$. For the comparison, the linear spring force is shown together with the nonlinear spring force*

The approach we demonstrate by [4.4] can be applied to a large number of different nonlinear restoring forces $\mathcal{F}(x)$, not necessarily odd with respect to x. In recent years, it has become possible to design and fabricate springs and supporting elastic elements for MEMS devices that have specific shapes (angled, curved, etc).. If we apply the Taylor expansion to the restoring force generated by these "designed" springs, it will contain all orders of x, even and odd:

$$F_{spring} \approx - \left(k_1x + k_2x^2 + k_3x^3 + k_4x^4...\right) \quad [4.6]$$

Thus, a more general model of a nonlinear resonator is:

$$\ddot{x} + 2\gamma\dot{x} + \omega_0^2 x + \sum_{i=2}^{N} \kappa_i x^i = F_{ext}/m \quad [4.7]$$

where we introduce the spring coefficients $\kappa_i = k_i/m$ that are normalized by the resonator mass m.

A nonlinear spring is termed *hardening* if, for large displacements, the restoring force of the spring will be greater than those of the linear spring. A nonlinear spring is termed *softening* if its restoring force decreases for larger displacements. For example, the Duffing (cubic) spring with $k_3 > 0$ is hardening because we add a term that will increase the magnitude of the restoring force if x increases. On the other hand, the Duffing (cubic) spring with $k_3 < 0$ is softening because we subtract a term and decrease the magnitude of the restoring force. Many springs have a more complex dependence of the restoring force on the displacement. They can display softening behavior for intermediate magnitudes of the displacement and hardening behavior for very large displacements. Sometimes, they are called locally softening but globally hardening springs.

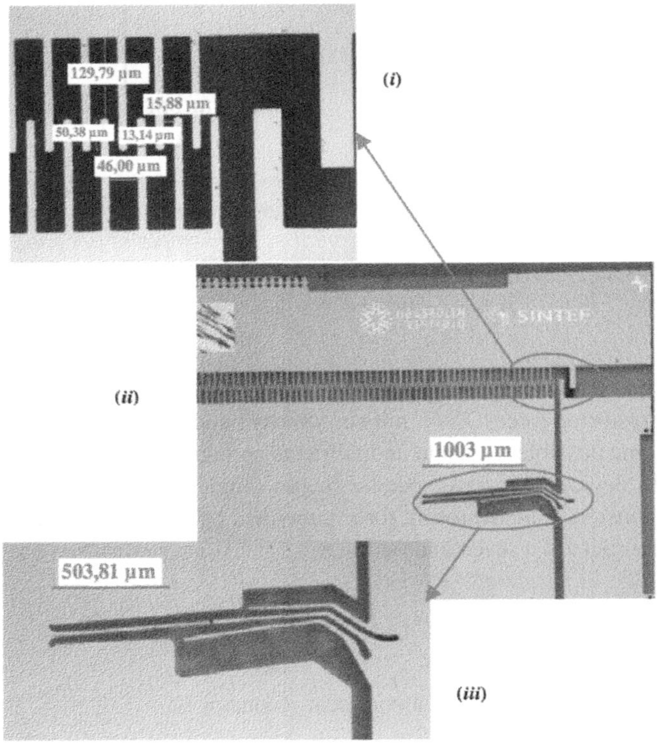

Figure 4.2. *Photograph of a MEMS resonator with a nonlinear spring showing strong softening effect. The photograph is taken from [NGU 10]*

There is a large number of examples of nonlinear microscale resonators [BEE 09, ZHU 10, LI 06, TAN 10, NGU 11, NGU 10, TVE 10, AMR 11] that are designed specifically to be nonlinear to improve the characteristics of harvesters. Nonlinear springs can be obtained by using angled, inclined springs or curved springs. For example, angled and curved spring of a particular configuration are found to display softening behavior as shown in Figure 4.2.

Finally, for the sake of completeness, we will discuss bistable modifications of resonators. We want to note here that these modifications are in particular common for macro- and microscale harvesters with piezoelectric transducers and are not commonly implemented with electrostatic transducers. For this reason, we will not go into much detail, but rather just describe it here. Such resonators are proposed for harvesting of energy from noise-like ambient motion [ERT 09, COT 09].

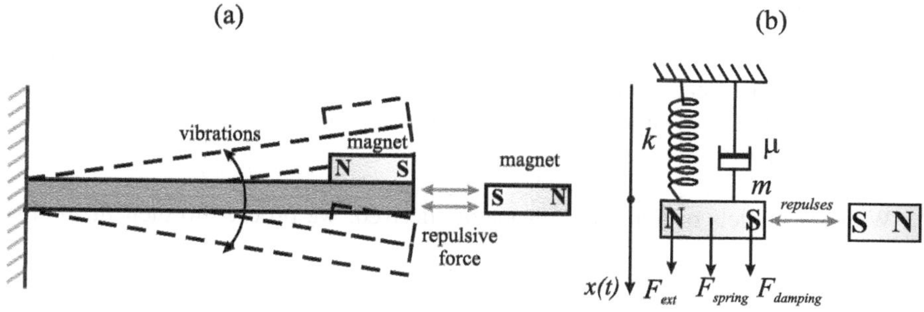

Figure 4.3. *Schematic of a bistable resonator*

Consider a system shown in Figure 4.3(a) or its 1D model in Figure 4.3(b). Assume for a moment that there is no magnet. In the absence of any external forces, the beam will stay still in equilibrium where the deflection is zero $x = 0$. Even if we add some small initial deflection to the beam position, the beam will eventually return to this equilibrium according to our 1D model solution [3.42]. Therefore, this equilibrium is stable.

If we assume that the tip of the beam is made of a magnetic material and if we place another magnet (note the location of the poles of the magnet) in the vicinity of the equilibrium, the two magnets will experience a repelling force. Even smallest variations in the beam's initial displacement or velocity (for example due to noise) will cause the beam to displace from the equilibrium at $x = 0$. Depending on the value of those small variations in the initial conditions, the beam will deflect up or down. Whatever force we apply to the beam, we will not be able to fix the beam in its "former" equilibrium point $x = 0$. We say that such an equilibrium becomes unstable.

We must assume that if the deflection is too large $x \gg 1$, the overall restoring force increases and tends to return the beam in the vicinity of the unstable equilibrium, otherwise the system would become globally unstable (we already know that realistic restoring forces are nonlinear functions of the displacement and so the magnitude of the restoring force in general increases with x). Thus, by introducing the magnet in the system as shown in Figure 4.3, we make the original equilibrium $x = 0$ unstable. But we also create two stable equilibrium positions for the beam (one up and one down). This is the reason why systems of this kind are called *bistable*.

The model equation that describes the simplest bistable resonator is the resonator of this form:

$$\ddot{x} + 2\gamma\dot{x} - \omega_0^2 x + \kappa x^3 = F_{ext}/m \qquad [4.8]$$

It seems to be very similar to [4.5] with the only difference that the term $\omega_0^2 x$ appears with the minus sign. This "minor" change significantly changes the behavior of such a resonator.

With the models of nonlinear resonators derived, we proceed to their qualitative study.

4.1.2. *Free oscillations of nonlinear resonator: qualitative description using potential wells*

We will adopt the same strategy as we used with linear resonators. Let us start by discussing the behavior of a nonlinear resonator with no input first to see what new effects are brought to resonators by mechanical nonlinearity.

A nonlinear restoring force, as we introduced it, depends only on the displacement of the resonator and therefore is a potential force. According to equation [3.19], we can write the potential energy of a nonlinear resonator:

$$\frac{dW_p}{dx} = m \cdot g(x) = -\mathcal{F}(x) \qquad [4.9]$$

For instance, we have already established that for a linear resonator its potential energy associated with the spring force is $W_p = kx^2/2$. Indeed, $F_{spring} = -kx$ and, as follows from the above equation, $W_p = -\int F(x)dx = \int kxdx = kx^2/2$.

Therefore, the potential energy of the (monostable) Duffing nonlinear resonator [4.5] and the bistable resonator [4.8] is:

$$W_p = \pm\frac{k_1 x^2}{2} + \frac{k_3 x^4}{4} \qquad [4.10]$$

The free unforced oscillations resulting from equations [3.6] and [4.3] (with $F_{ext} = 0$) can be seen as the coordinates of a particle that moves (oscillates) on a curved surface whose profile has the shape of the potential function W_p. If such a surface is ideal (no friction), it will correspond to the case $\gamma = 0$, and the particle slides (oscillates) on this surface without loss of energy. This analogy is shown in

Figure 4.4 where a particle of mass m moves in a potential well W_p. The displacement and the velocity of such a particle is the same as x and v we obtain from solving differential equations [3.6] and [4.3].

Figure 4.4 shows the potential functions, W_p, of a linear, Duffing nonlinear and bistable resonators. For the linear resonator, the potential function $W_p = kx^2/2$ is a parabola. For the nonlinear resonator with $W_p = \alpha x^2 + \beta x^4$ ($\alpha = k_1/2 > 0$ and $\beta = k_3/4 > 0$), the potential function resembles a parabola. Although it is qualitatively similar to the linear case, the large-scale oscillations of the linear and nonlinear resonators will be different because the curvature of the potential wells for these two resonators are very different. Only small-scale oscillations around the local minimum of W_p (around $x_{eq} = 0$) will be similar in both systems. For a nonlinear bistable resonator shown in Figure 4.4(c), the potential function is given by $W_p = -\alpha x^2 + \beta x^4$ ($\alpha = k_1/2 > 0$ and $\beta = k_3/4 > 0$). Such a potential well has two local minima and one local maximum. Depending on initial conditions, the particle, as shown Figure 4.4(c), can oscillate either around one minimum of W_p or the other. Each minimum corresponds to one of the two stable equilibria of the resonator [4.8] we discussed in the previous section. This potential is termed *bistable*, or, in the most general case, *multistable*.

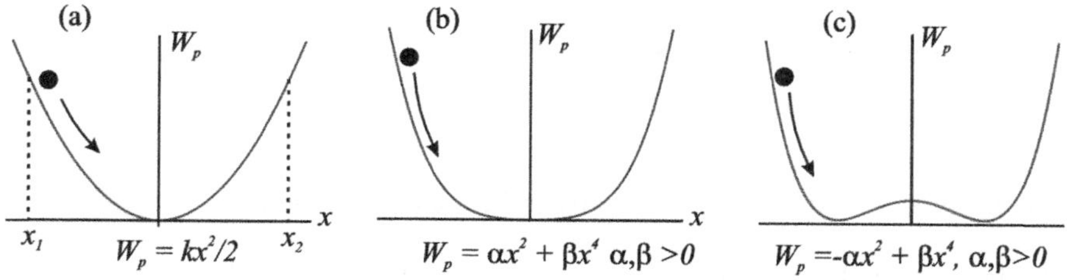

Figure 4.4. *Schematic view of the potential energy W_p of three different resonators; a) Linear resonator with $g(x) = \omega_0^2 x$ and $W_p = kx^2/2$; b) Nonlinear resonator with $g(x) = \omega_0^2 x + \kappa x^3$ and $W_p = \alpha x^2 + \beta x^4$ where $\alpha = k_1/2$, $\beta = k_3/4$ and $\alpha, \beta > 0$; c) Nonlinear resonator with $g(x) = -\omega_0^2 x + \kappa x^3$ and $W_p = -\alpha x^2 + \beta x^4$ where $\alpha = k_1/2$, $\beta = k_3/4$ and $\alpha, \beta > 0$*

From our study of the linear resonator, we know that in the absence of an external force, the linear resonator will display decaying oscillatory process at a frequency that is close to the natural frequency of the resonator ω_0. We could anticipate similar behavior in the nonlinear case as well. For a start, it is very easy to show that the energy relation for the nonlinear resonator is exactly the same as that of the linear resonator:

$$\frac{d}{dt}\left[\frac{m\dot{x}^2}{2} + W_p\right] = -\mu\dot{x}^2 \quad [4.11]$$

The potential function W_p can now be some other function of the displacement, different from the parabolic $kx^2/2$ expression of the linear resonator. For a nonlinear resonator, the same as for a linear resonator, in the absence of friction $\gamma = 0$, its energy is conserved. While in the presence of friction $\gamma > 0$, energy dissipates. Thus, for a realistic nonlinear system with $\gamma \neq 0$, all energy that was initially in the resonator will be dissipated and all motion will stop. In this, both systems are similar.

4.1.3. *Free oscillations of nonlinear resonator: semi-analytical approach*

There is a substantial difficulty in the analysis of nonlinear resonators. It is caused by the fact that most nonlinear equations do not have closed-form solutions [NAY 93, NAY 08b]. We are not able to write the solution to [4.3] or to [4.5] even in case when $F_{ext} = 0$. However, a range of different semi-analytical methods exist for nonlinear systems. Their application in the context of a kinetic energy harvester (KEH) will be discussed in Chapter 5. Omitting the derivation in this chapter (we refer you to Chapter 5), we provide below the semi-analytical solution to equation [4.5] that describes free oscillations (decaying oscillatory process) of the Duffing resonator with no external force:

$$x(t) \approx a_0 e^{-\gamma t} \cos\left(\omega_0 t + \varphi_0 + \frac{3\kappa a_0^2}{16\omega_0\gamma}\left(1 - e^{-2\gamma t}\right)\right) + \frac{\kappa a_0^3 e^{-3\gamma t}}{32\omega_0^2} \cos\left(3\omega_0 t + 3\varphi_0 + \frac{9\kappa a_0^2}{16\omega_0\gamma}\left(1 - e^{-2\gamma t}\right)\right) \quad [4.12]$$

where a_0 and φ_0 are the quantities that describe the initial conditions.

By inspecting [4.12], we see that the solution to a free, unforced Duffing resonator contains not only the fundamental component $\cos(\tau + \varphi)$ but also the third harmonics given by the term $\cos(3\tau + 3\varphi)$. We say that such oscillations are *anharmonic*. This is the case for all nonlinear resonators – their oscillations contain a number of various harmonics (however, the amplitude of each harmonic decreases with the increase in its number).

Figure 4.5 shows the evolution of the displacement x of the Duffing equation [4.5] as a function of time t at $\omega_0 = 1$, $\gamma = 0.05$ and $\kappa = 0.5$ obtained from a numerical simulation. For comparison, we also give the solution of a linear resonator at the same dissipation $\gamma = 0.05$ (and $\kappa = 0$ in this case) and the approximated solution [4.12] for comparison.

The comparison of the oscillations given in Figure 4.5 also reveals that the frequency of decaying free oscillations of the Duffing resonator is different from the frequency of the linear resonator (all parameters are the same but nonlinearity is absent). This can be seen by observing the moments of time when both solutions cross the horizontal axis – it clearly occurs at different instants.

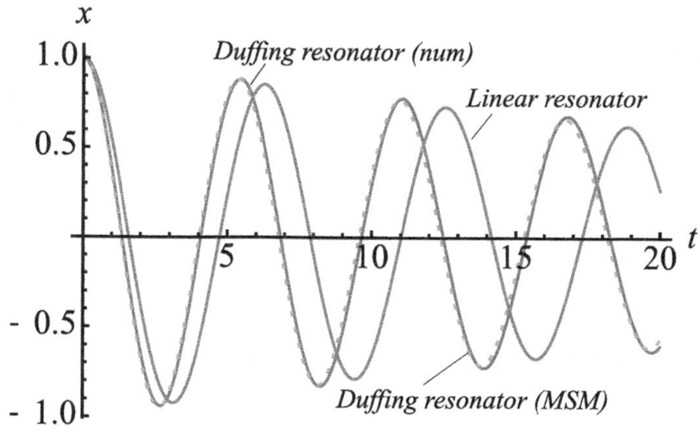

Figure 4.5. *Evolution of the displacement x as a function of time for a linear resonator* [3.6] *and the Duffing resonator* [4.5] *obtained from numerical simulation of the equation. For the comparison, the approximated solution* [4.12] *obtained from the multiple scales method is shown by the dashed line.* $\omega_0 = 1$, $\kappa = 0.5$, $\gamma = 0.05$. *For a color version of the figure, see www.iste.co.uk/basset/keh.zip*

This is surprising. We know that the frequency of free oscillations of a linear resonator is entirely defined by the constant parameter ω_0 and is the same no matter what initial conditions we take. The natural frequency ω_0 is a fundamental parameter of the linear resonator. However, it appears that there is no constant natural frequency that can be associated with a nonlinear resonator. Its frequency depends on the amplitude of oscillations in the resonator and on initial conditions. We say that such oscillations are *anisochronous*. In contrast, the oscillations of a linear resonator would be called *isochronous*. The dependence of the frequency of nonlinear oscillations on the amplitude (energy) explains the fundamental phenomenon of nonlinear resonance. We will return to this in Chapter 5 for a more detailed discussion.

4.1.4. *Forced nonlinear resonator and nonlinear resonance*

We have already studied a linear resonator under external harmonic actuation and established that if the frequency of the external force is close to the natural frequency of the resonator, we observe resonance. In a nonlinear resonator, as we know from the previous section, oscillations are anisochronous, i.e. the frequency of such oscillation is not a constant, but rather depends on the energy of the resonator. Resonance is also possible in nonlinear systems, but the resulting forced oscillations are more complex than in the linear case. We will consider a Duffing resonator as an example to illustrate nonlinear resonance.

The steady-state response of the Duffing resonator [4.5] to the external harmonic force (divided by the mass m) $F_{ext} = F_0 \cos(\omega_{ext} t + \phi_0) = A_{ext} \cos(\omega_{ext} t + \phi_0)$ will be derived in Chapter 6 as it requires a knowledge of methods of nonlinear analysis that we will overview in that chapter. Below, we will give the semi-analytical expressions for this response:

$$x(t) \approx a_0 \cos(\omega_{ext} t + \phi_0 - \psi_0) + \frac{\kappa\, a_0^3}{32\omega_0^2} \cos(3\omega_{ext} t + 3\phi_0 - 3\psi_0) \qquad [4.13]$$

where a_0 and ψ_0 is found from the following equations:

$$\frac{F_0^2}{4\,(\omega_0)^2} = (\gamma a_0)^2 + \left(\sigma\,\omega_0 a_0 - \frac{3\kappa a_0^3}{8\,\omega_0}\right)^2 \qquad [4.14]$$

and

$$\frac{F_0}{2\omega_0} \sin\psi_0 = \gamma a_0, \qquad \frac{F_0}{2\omega_0} \cos\psi_0 = -\sigma\,\omega_0\, a_0 + \frac{3\kappa a_0^3}{8\,\omega_0} \qquad [4.15]$$

where we introduced the normalized frequency mismatch between the resonator linear frequency ω_0 and the frequency of the external driving force ω_{ext}: $\sigma = \omega_{ext}/\omega_0 - 1$.

If you recall the case of resonance in the linear resonator (linear resonance), the amplitude of forced oscillations was given by equation [3.52]. It was a simple algebraic expression, also called the linear resonance curve, that allowed us to find this amplitude if the amplitude of the external force F_0 and its frequency ω_{ext} were given. Note that there was only one unique value of the amplitude in linear resonance.

Now, if we want to find the amplitude a_0 of forced oscillations in a nonlinear resonator, we must solve a nonlinear equation [4.14]. This equation is called the *nonlinear resonance curve*. Since equation [4.14] is a polynomial equation of the sixth order in terms of a_0, depending on the parameters F_0, σ, ω_0, κ and m, we may obtain either one real positive value of a_0 or three real positive values of a_0 that satisfy this equation[1]. However, the result is again surprising – the amplitude a_0 is not unique.

1 The equation of this form will have six roots in general. However, they will appear as pairs $\pm a_{0,1}$, $\pm a_{0,2}$ and $\pm a_{0,3}$. We are interested only in amplitudes a_0 that are positive: $a_0 > 0$. Thus, there can be either one such amplitude $a_{0,1}$ (with $a_{0,2}$ and $a_{0,3}$ being complex conjugate) of three of them $a_{0,\{1,2,3\}}$.

Let us plot a figure, similar to the response shown in Figure 3.5. The amplitude of steady-state forced oscillations in the Duffing resonator a_0 as a function of the frequency mismatch $\sigma = \omega_{ext}/\omega_0 - 1$ obtained from [4.14] is shown in Figure 4.6. Three cases are shown: $\kappa = 0$ (corresponds to the resonance in a linear damped resonator), $\kappa < 0$ (nonlinear, spring softening) and $\kappa > 0$ (nonlinear, spring hardening). All of them are calculated for the same amplitude of the external force F_0. In contrast to the linear case shown by the gray line, resonance in nonlinear resonators leads to the bistability of the nonlinear resonance curve, which we will call *bimodality*, in this book[2], and *hysteresis*. Bimodality appears because we can obtain three different solutions for a_0 from the nonlinear resonance curve [4.14].

From this figure, we can see that for certain values of the frequency mismatch σ there are three coexisting branches of the graph. This means that there are three solutions for a_0 found from [4.14]. The further analysis of these solutions shows that two of them are stable (shown by the solid line) while the third is not stable (this branch is shown by the dashed line). Depending on initial conditions $x(t_0)$ and $v(t_0)$, the resonator can display forced vibrations with a large amplitude $a_{0,1}$ or with a smaller amplitude $a_{0,2}$. This phenomenon is called bimodality (multimodality).

Now assume that we can slowly change the frequency of the external oscillations ω_{ext}, for example over the range from some $-\omega^*_{ext}$ to $+\omega^*_{ext}$ (this is called a frequency sweep). If we sweep the frequency of the external force in the direction shown by the arrows on the resonance curves, we will observe hysteresis. Let us consider the case $\kappa > 0$, for example. Increasing the frequency ω_{ext}, we will observe a gradual increase in the oscillation amplitude (arrow moving right in the figure). When the peak of the resonance curve is reached at some $\omega_{ext,1}$, we will observe a sharp jump as we will switch to the lower stable branch of the curve. Decreasing the frequency ω_{ext}, we also will observe gradual increase in the forces oscillations. At another frequency $\omega_{ext,2} < \omega_{ext,1}$, we observe a jump to a higher stable branch of oscillations. This phenomenon is known as the hysteresis of the nonlinear resonance curve.

As is seen from Figure 4.6, nonlinear resonators, in principle, can respond to external vibrations (maintain resonance, i.e. large amplitude oscillations) for a wide frequency range of ω_{ext}. However, when we drive a nonlinear resonator to nonlinear

2 In the theory of resonators and oscillators, it is common to call the bistability of the resonance curve simply as bistability. This means that there are two coexisting stable regimes of forced oscillations in a nonlinear resonator, one with an amplitude $a_{0,1}$ and the other with an amplitude $a_{0,2}$. At the same time, the original potential function (see formula [4.10] with the plus sign in front of the $k_1x^2/2$ term, also shown in Figure 4.4b) of the resonator that displays this resonance response (in our case, the Duffing resonator [4.5]) is not bistable. The reader should be aware that the bi-stability of the potential function and the bistability of the nonlinear resonance curve are two different and independent phenomena. In order to avoid any confusion, we will term the bistability of the nonlinear resonance curve as bimodality (or multimodality in the most general case) throughout the book.

resonance in the region of the large amplitude response and hysteresis, we must remember about its bimodality. Since there are two coexisting stable regimes (one with a large amplitude and the other with a small amplitude), we risk jumping to another stable branch that results in small-amplitude vibrations. Thus, we must ensure that the resonator is kept at the upper branch of the bimodal curve as shown in Figure 4.6. In the context of energy harvesting, this may limit the practical implementation of nonlinear resonance.

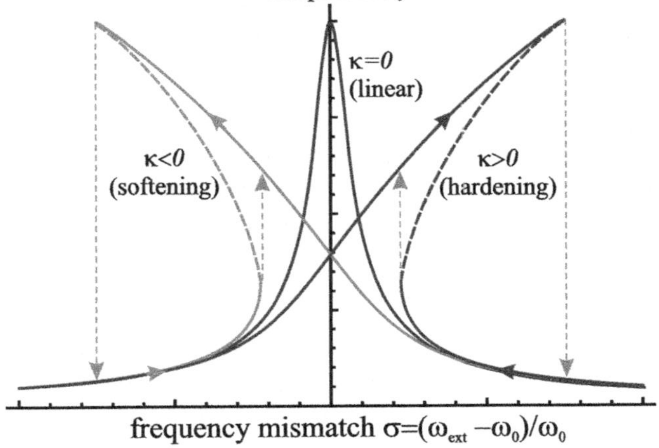

Figure 4.6. *Resonance in linear and nonlinear resonators. The amplitude a_0 of steady-state forced oscillations as a function of the normalized frequency mismatch $\sigma = \omega_{ext}/\omega_0 - 1$. The light gray line corresponds to the linear case when $\kappa = 0$ (see also Figure 3.5), while the blue and green lines correspond to nonlinear cases ($\kappa > 0$ and $\kappa < 0$). For the nonlinear cases, we can see multimodality and hysteresis: there is a segment of the resonance curve $a_0(\omega_{ext})$ where two stable solutions of a_0 and one unstable a_0 (shown by the dashed line) coexist. For a color version of the figure, see www.iste.co.uk/basset/keh.zip*

This range of ω_{ext}, for which a nonlinear resonator can effectively respond to external vibrations, can be significantly larger than the resonance frequency range of a linear resonator. The qualitative comparison of the high-Q linear and nonlinear resonators is shown in Figure 4.7. You can compare it with the expression for the linear bandwidth BW and Figure 3.5. A very qualitative explanation of nonlinear resonance is based on the fact that nonlinear oscillations are anisochronous. Their "natural" frequency is not constant and changes with the amplitude of vibrations. The larger the amplitude a_0, the larger the change in the resonance frequency and the wider the frequency response of a nonlinear resonator. This is a particularly important advantage of nonlinear systems over linear ones in the context of energy harvesting systems. In addition, this behavior is universal for nonlinear systems: it appears not only in the Duffing resonator, but also in other nonlinear resonator with polynomial nonlinearities.

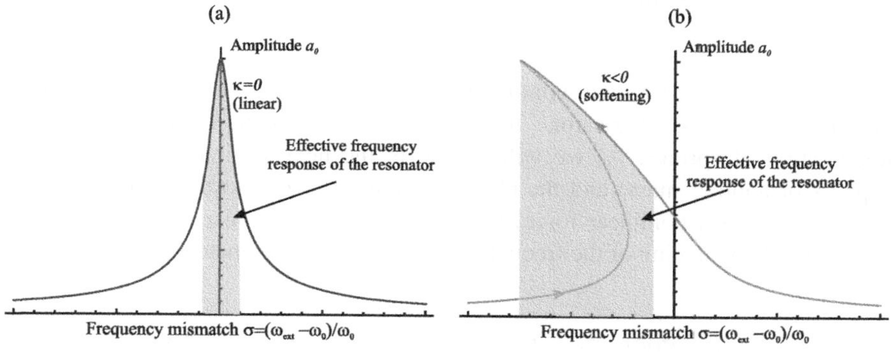

Figure 4.7. *Resonance in linear and nonlinear resonators. The high-Q linear resonator displays a narrow resonance peak near the natural frequency ω_0. The nonlinear resonator displays a wider peak near a frequency $\omega^* < \omega_0$ (for the softening case $\kappa < 0$). The effective bandwidth (of the frequency response) of the linear resonator is lesser than that of the nonlinear resonator. $BW_{lin} < BW_{nonlin}$. In this example, the bandwidth is taken at the level where the amplitude of the resonator displacement decreases by two compared to the peak value*

Suppose that we have to compare two energy harvesting systems: one using a high-Q linear resonator and the other with a high-Q nonlinear resonator with mechanical nonlinearity represented by a polynomial force as in equation [4.7]. If we drive the harvester by a harmonic vibration (or vibrations) at a frequency (or frequencies) $\omega_{ext} \approx \omega_0$, we will obtain large amplitude vibrations in the linear harvester. The characteristic of these vibrations are completely predictable. Now, if for some reason the frequency of the external vibrations ω_{ext} changes and shifts from ω_0, the linear resonator will be quickly "detuned" from resonance and the amplitude of vibration will drop significantly. On the other hand, the nonlinear resonator will continue to be "in resonance" (recall, however, again about its bimodality).

Researchers and KEH designers seek to expand the resonance response of the mechanical part of a harvester (keeping the Q-factor of the resonator large) – this is a desirable effect [ERT 09, BAR 12, STA 10, DAQ 10, AND 10, ZHU 10, MIK 10]. This is only possible by using some sort of nonlinearity. One possibility is to introduce mechanically induced nonlinearities to the resonators. As we have just learnt, these nonlinearities lead to one of the two fundamental nonlinear effects – the spring hardening effect or the spring softening effect. The spring hardening effect is observed when the effective resonance frequency increases with the amplitude of external vibrations and the amplitude of the resonator vibrations. The spring softening effect is observed when the effective frequency decreases with the amplitude of external vibrations and the amplitude of the resonator vibrations. We will remind the reader that these cases are shown in Figure 4.6 and are manifested through the "tilt" of the resonance curve to the right (hardening) or to the left (softening).

However, the same two nonlinear effects (hardening or softening) can be introduced into the resonator, using mechanisms other than the nonlinear modification of springs. As we will review in the next section, the impact (with stoppers or another mass) and the electric nonlinearity due to the transducer force (recall that F_t is a nonlinear force) will bring similar nonlinear effects and can be utilized for the widening of the frequency response of the resonator.

4.2. Review of other nonlinearities affecting the dynamics of the resonator: impact, velocity and frequency amplification and electrical softening

As we mentioned, the wide frequency response of the mechanical resonator in a KEH is a desirable effect and in practice it can be achieved by mechanical nonlinearities or other mechanisms.

[COT 14b] introduces a multiple mass system and demonstrates how the dynamics of multiple mass systems can be used to significantly enhance the power output of KEHs compared to single-mass KEH designs. The particular configuration of such a system with a two-mass resonator is shown in Figure 4.8. The reason why this structure was proposed for energy harvesting is that it allows us to explore the velocity amplification principle through the impact of the proof mass with another mass. Indeed, note that the second mass is detached from the lower (heavier) mass and experiences an impact with the lower mass if the system is driven by large amplitude vibrations. To explain the principle of velocity amplification, consider an elastic collision of the two travelling masses. The conservation of total momentum dictates that the final velocity of the smaller mass v_{2f} is equal to

$$v_{2f} = \frac{(e+1)m_1 v_{1i} + (m_2 - e\, m_1)v_{2i}}{m_1 + m_2} \qquad [4.16]$$

where e is the coefficient of where restitution and v_{1i} and v_{2i} are the velocities of the first and second masses before the impact. By assuming that $e = 1$ and $m_1/m_2 \to \infty$, we can obtain the upper limit to the velocity of the second mass $v_{2f} = 2v_{1i} - v_{2i}$. In the case of equal but opposite velocities $v_{1i} = -v_{2i}$, we can obtain a substantial gain in the velocity of the second mass $v_{2f} = 3v_{2i}$. This is particularly important for KEHs whose transduction mechanism is proportional to the velocity. This is the case for electromagnetic harvesters where the induced electromotive force (e.m.f) in the coil connected to the resonator is directly proportional to the velocity of the second mass v_2.

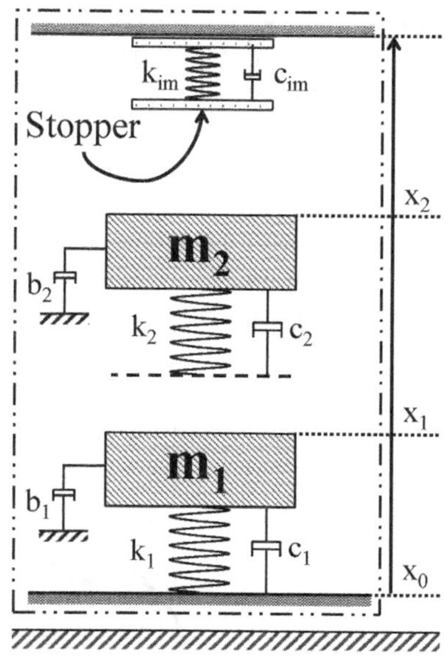

Figure 4.8. *Schematic structure of a two-mass resonator for wideband energy harvesting using the velocity amplification principle from [COT 14b, FU 14]*

However, the impact dynamics of the two-mass systems introduce nonlinearity and cause an additional effect – widening of the frequency response of the system. In [FU 14], it has been shown that the dynamics of the two masses are governed by piecewise-defined differential equations. Therefore, this system as a whole cannot be considered as a linear system. It displays nonlinear and chaotic dynamics over a broad range of controlling parameters (such as A_{ext} and f_{ext}). Figure 4.9 illustrates a typical response of such a system to a frequency sweep (the parameters of the device can be found in [FU 14]). Note that since the displacement of the first and the second mass is irregular, the frequency response is given in terms of the root mean square of the power generated by the harvester (the generated power is linearly proportional to the velocity of the second mass). Also note that the bandwidth of the response is rather wide compared to a linear resonator case and increases with A_{ext}.

The study by Cottone *et al.* [COT 14a] also uses impacts to widen the frequency response of electrostatic energy harvesters. In this chapter, the authors suggest a new structure of the MEMS resonators for an electrostatic kinetic energy harvester (e-KEH) that exploits mechanical velocity amplification by multiple mass impacts in combination with elastic stoppers (see Figure 4.10). When the system is shaken at low frequency in the range 10–60 Hz, a miniature ball, placed in the cavity of the MEMS proof mass, impacts with it and transfers kinetic energy to the gap-closing comb transducer. The transducer, however, resonates at its natural frequency, which is higher than that of the micro ball. The natural frequency of the transducer/resonator is typical for MEMS devices and in the range of over 100 Hz. In addition, elastic stoppers amplify the proof mass and ball velocity through collisions with the fixed frame. The system combines the advantages of impact nonlinearity (wider frequency response) with the frequency-up conversion mechanism [GAL 11b]. Such a system can be driven by vibrations at a lower frequency (more common for mechanical vibration and much lowers than the natural frequency of the MEMS resonator) and it will respond to them as shown in that work.

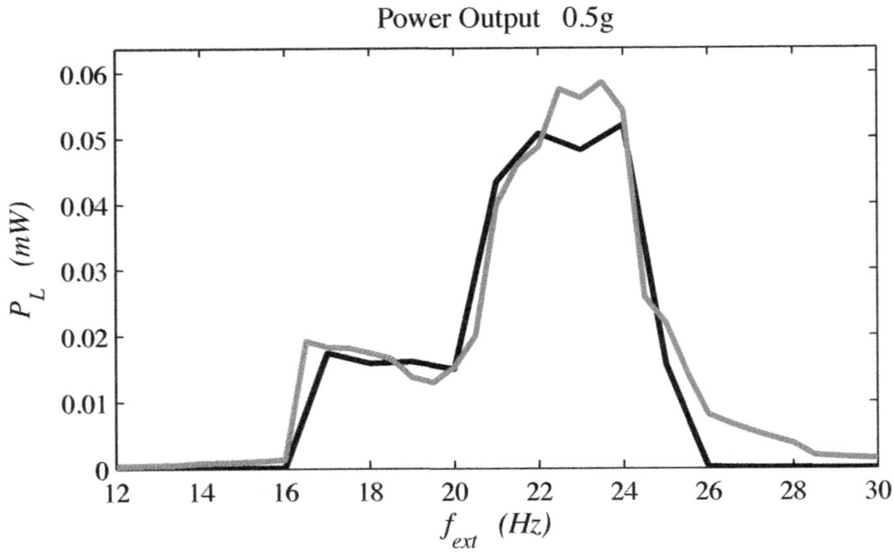

Figure 4.9. *Root mean square (RMS) of the converted power P_L versus the frequency of the input vibrations ω_{ext} of an electromagnetic harvesters utilizing the two-mass resonator with velocity amplification. The details of the system can be found in [FU 14]. The black line is the experimental measurement while the red line is the model proposed in the latter study. Note the wideband response of the system, i.e. the absence of clear and narrow peak around the resonance frequency $f_{res,1} = 23.1$ Hz*

The effect of stoppers on the performance of electrostatic harvesters has been exploited in studies [LE 12, TAK 15]. In these works, the authors show using simulation and experimental results that the impact with stoppers has positive effects.

It increases the bandwidth of the resonator and increases the converted power for large acceleration power spectral densities.

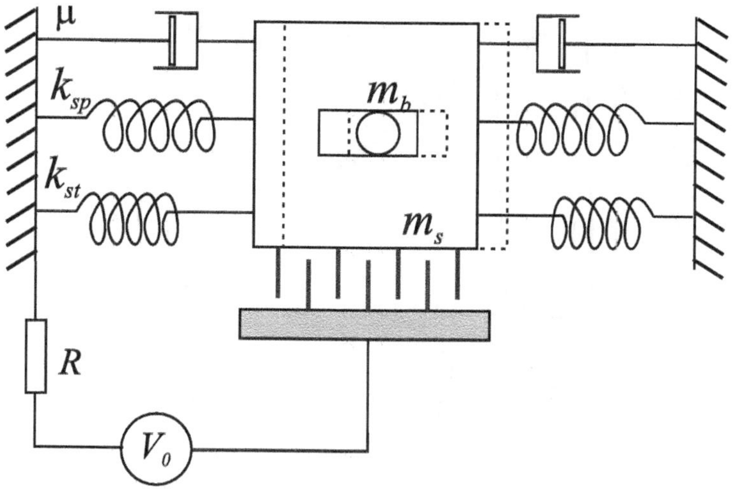

Figure 4.10. *Mechanical structure of the MEMS resonator using impacts as suggested in study [COT 14a]*

Finally, the transducer force F_t (nonlinear as we discussed at the beginning of this chapter) causes similar effects on the resonator as those produced by nonlinear mechanical forces. For instance, the transducer force causes hysteresis, multistability, and so-called *electrostatic softening* (or in some cases – *electrostatic hardening*) effects even in the absence of mechanical nonlinearity (compare Figure 4.11 with Figure 4.6). In the context of electrostatic harvesters, these effects are quantified, described and measured in studies [BAS 14, GAL 15, O'RI 15]. In particular, in the study by O'Riordan *et al.* [O'RI 15] the authors treat the transducer force in a way similar to nonlinear mechanical forces and provide a theory to explain and quantify the electromechanical coupling due to this force. We will discuss this in Chapter 6.

4.3. Concluding remarks: effectiveness of linear and nonlinear resonators

We will briefly summarize the advantages and disadvantages of linear and nonlinear resonators. As we have just learnt, mechanical and non-mechanical nonlinearities cause similar effects to the resonator of a KEH. The most "obvious" and useful effect is the widening of the frequency response of the resonator with respect to the frequency of external vibrations ω_{ext}. This is clearly seen from Figure 4.7 and is demonstrated in many studies of nonlinearity in energy harvesters.

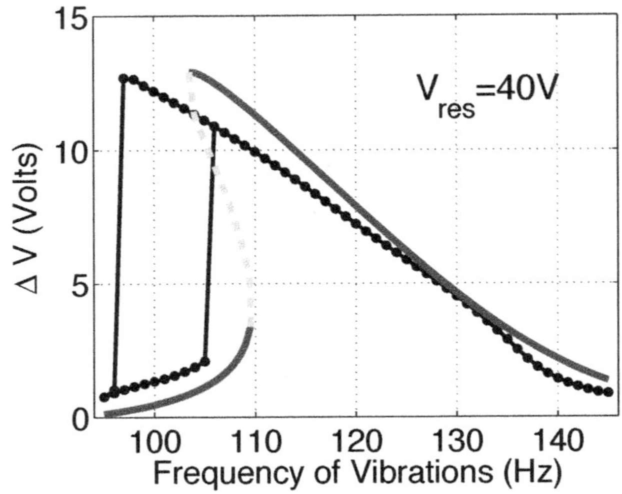

Figure 4.11. *Nonlinear effects that are caused by electrostatic transducer force F_t are similar to those caused by mechanical nonlinear forces. This figure shows the effect of hysteresis and frequency shift due to F_t (compare with Figure 4.6). The two curves correspond to a theory and measurements whose details can be found in the study [O'RI 15]. For a color version of the figure, see www.iste.co.uk/basset/keh.zip*

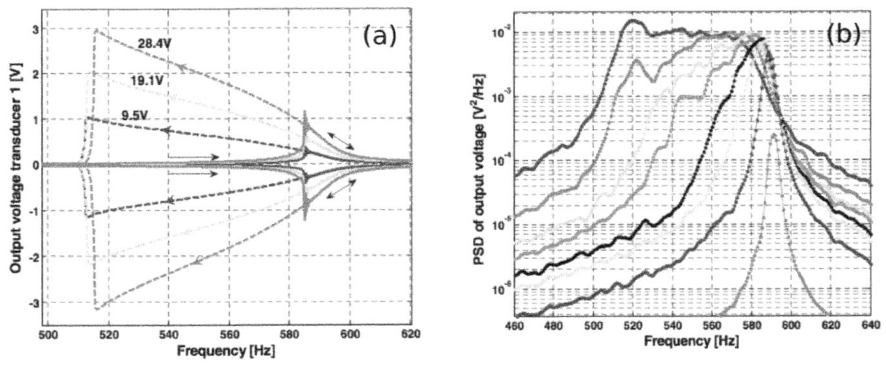

Figure 4.12. *a) Frequency sweep of a nonlinear resonator for KEHs showing strong softening behavior due to mechanical nonlinearities; b) Excitation of this KEH by wideband noise-like vibrations showing the response of the KEH over a wider range of frequencies. Figures are taken from [NGU 10]. For a color version of the figure, see www.iste.co.uk/basset/keh.zip*

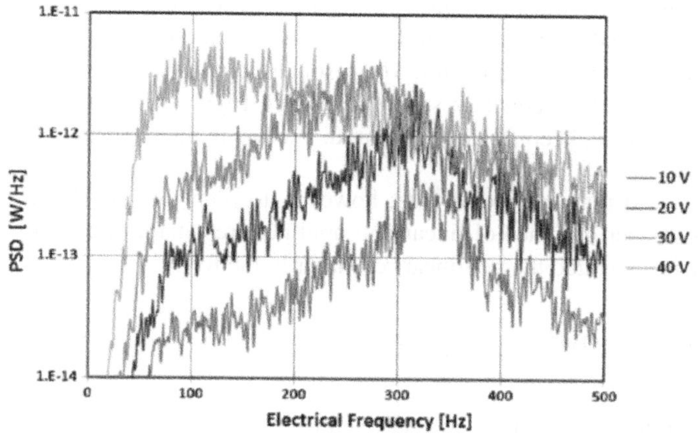

Figure 4.13. *Excitation of a KEH using electrostatic and impact nonlinearities by wideband noise-like vibrations. Note that these nonlinearities cause a similar effect as that of mechanical nonlinearities. The KEH displays a wide frequency response. Figure is taken from [BAS 14]. For a color version of the figure, see www.iste.co.uk/basset/keh.zip*

Additionally, nonlinearities can shift (modify) the resonance frequency and may cause hysteresis (in the case of strong nonlinearities).

Therefore, for linear resonators:

– *disadvantage*: narrow bandwidth and high frequency selectivity;

– *advantage*: unique amplitude and stability.

While for nonlinear resonators:

– *disadvantage*: multimodality and hysteresis may lead to jumps between two or more possible solutions;

– *advantage*: wider bandwidth and less selective frequency response.

Nonlinear resonators seem to be particularly suitable for wideband energy harvesting and noise harvesting (i.e. when external vibrations represent noise) due to their wider frequency response. This property has been exploited in KEHs by designing and fabricating MEMS resonators with nonlinear springs and other nonlinearities that we discussed in the earlier section. It has been shown that such nonlinear systems respond more effectively to ambient wideband or noise-like vibrations. For example, the study by Nguyen *et al.* [NGU 10] shows that MEMS energy harvesters with nonlinear springs showing that strong softening effects (similar to the one shown in Figure 4.6 for $\kappa < 0$) respond very effectively to noise actuation. Figure 4.12 shows the nonlinear behavior and response to wideband noise vibrations of a nonlinear resonator. The resonator displays strong softening behavior as follows from the frequency sweep shown in Figure 4.12(a) and responds very effectively to noise, keeping large amplitude motion in the resonator (Figure 4.12(b)).

Having said that, we would like, however, to point out [HAL 13], a study that compares the performance of linear and nonlinear resonators driven by white noise. The study investigated the nonlinear resonator in the form:

$$m\ddot{x} + \mu\dot{x} + W_p(x) = m\xi + F_t \qquad [4.17]$$

where the transducer force F_t is taken into account and the harvester resonator is actuated by white noise ξ. The resonator is considered in a general nonlinear form with the elastic potential W_p representing mechanical nonlinearities. The study by Halvorsen [HAL 13] uses the linear (linearized) transducer force F_t that describes piezoelectric harvesters and some particular cases of electrostatic harvesters. The important conclusion given by the author is that nonlinear harvesters are not fundamentally better than linear harvestors. A rigorous upper bound on the output power for arbitrary elastic potential W_p is derived in this chapter. With this, it is possible to find a linear device that performs equally well as the upper bound. It can be concluded that the use of nonlinear springs and nonlinear potentials is dictated by necessity (such as the presence of physical or implementation constraints) rather than any advantage in performance.

Finally, let us refer a reader who is interested in obtaining a deeper knowledge on linear and nonlinear resonators to numerous publications on this subject [AND 87, RAB 89, NEJ 92, NAY 08b, TRU 01, KUZ 02].

5

Fundamental Effects of Nonlinearity

In Chapters 3 and 4 we discussed the behavior of linear (in detail) and nonlinear (briefly) resonators. We saw that in the case of nonlinear resonators, there are a number of effects caused by nonlinearities (spring softening, hardening and hysteresis). This chapter aims to provide a more detailed discussion on these nonlinear effects. We overview a number of fundamental effects that are caused by nonlinearities (mechanical and electrical) and methods that can be used to analyze nonlinear systems in general.

5.1. Fundamental nonlinear effects: anisochronous and anharmonic oscillations

In Chapter 4, we introduced the equation of the Duffing resonator:

$$\ddot{x} + 2\gamma\dot{x} + \omega_0^2 x + \kappa x^3 = F_{ext}/m \tag{5.1}$$

and gave (without deriving it) the following semi-analytical expression:

$$\begin{aligned} x(t) &\approx a_0 e^{-\gamma t}\cos\left(\omega_0 t + \varphi_0 + \frac{3\kappa a_0^2}{16\omega_0\gamma}\left(1 - e^{-2\gamma t}\right)\right) \\ &+ \frac{\kappa a_0^3 e^{-3\gamma t}}{32\omega_0^2}\cos\left(3\omega_0 t + 3\varphi_0 + \frac{9\kappa a_0^2}{16\omega_0\gamma}\left(1 - e^{-2\gamma t}\right)\right) \end{aligned} \tag{5.2}$$

that describes free oscillations of the Duffing resonator (in this chapter, we will derive it). In equation [5.2], a_0 and φ_0 are quantities related to the initial conditions.

By inspecting [5.2], we first see that the solution to a free, unforced Duffing resonator contains not only the fundamental component $\cos(\omega_0 t + \varphi)$ but also the third harmonic given by the term $\cos(3\omega_0 t + 3\varphi)$. Oscillations that contain harmonics of the fundamental frequency are *anharmonic*, and therefore the oscillations of the Duffing resonator are anharmonic too. In general, this is the case for all nonlinear resonators: their oscillations contain a number of various harmonics, whose spectrum is defined by the particular form of the nonlinearity.

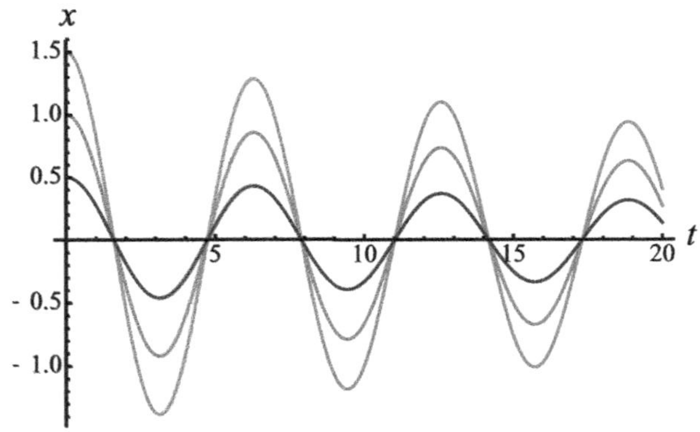

Figure 5.1. *Natural frequency of the linear resonator. The frequency of its eigen (free) oscillations does not depend on the initial state of the resonator: see how all three waveforms cross the horizontal axis of time at the same instants. This implies that the period of all three oscillation waveforms of the linear resonator is the same;* $\omega_0 = 1$ *and* $\gamma = 0.1$. *For a color version of the figure, see www.iste.co.uk/basset/keh.zip*

Second, expression [5.2] does not contain a term that could serve as the natural frequency of the oscillation. Let us look at another example. Figure 5.1 shows three solutions of the linear resonator that start from different initial conditions, $x_0^{(1)} = 0.5$, $x_0^{(2)} = 1.0$ and $x_0^{(3)} = 1.5$. All three waveforms cross the horizontal axis of time at the same instants. This implies that the period of all three oscillation waveforms of the linear resonator is the same. On the other hand, Figure 5.2 shows three solutions of the Duffing resonator [5.1] that also start from different initial conditions. The comparison of these three oscillations reveals that each of them decays at a different frequency. This can be seen by observing the instants of time when the three solutions cross the horizontal axis (it clearly occurs at different instants).

We conclude that there is no "natural" frequency that can be associated with a nonlinear resonator. Its frequency depends on the initial conditions (equivalent to saying that it depends on the amplitude of resonator oscillations or on the energy stored in the resonator). We say that such oscillations are *anisochronous*. In contrast, the oscillations of a linear resonator would be called isochronous. The dependence of the frequency of nonlinear oscillations on the energy of the resonator explains the fundamental phenomenon of nonlinear resonance.

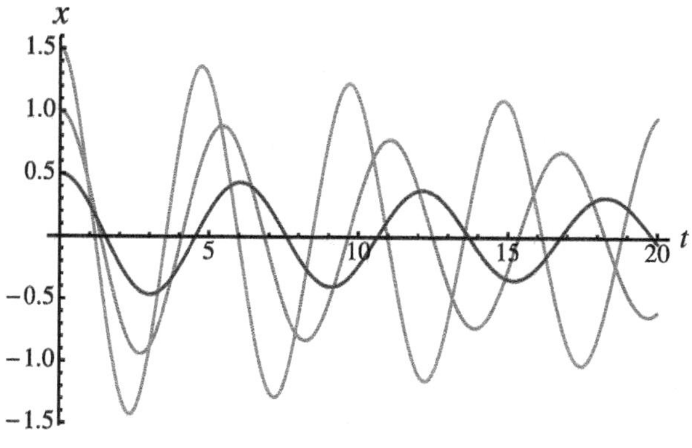

Figure 5.2. *The absence of a "fixed" natural frequency of the nonlinear Duffing resonator. The frequency of its free oscillations depends on the initial state of the resonator. The three waveforms start at different initial conditions and cross the horizontal axis of time at different instants. This implies that the frequency of these oscillation waveforms is different;* $\omega_0 = 1$, $\gamma = 0.1$ *and* $\kappa = 0.1$. *For a color version of the figure, see www.iste.co.uk/basset/keh.zip*

We have mentioned that the oscillations resulting from [5.1] can be seen as the coordinates of a particle that moves (oscillates) on a curved surface whose profile has the shape of the potential function W_p. We can recall that $F(x) = -dW_p/dx$. If such a surface is ideal (no friction), it will correspond to the case $\gamma = 0$. The particle slides (oscillates) on this surface without loss of energy. This analogy is shown in Figure 4.4 where a particle of mass m moves in a potential well W_p. The displacement and the velocity of such a particle is the same as x and v, which we obtain from solving the differential equations of linear or nonlinear resonators with the relevant potential well W_p.

Let us use this analogy to investigate (at a qualitative level) the period and the frequency of the nonlinear resonator as a function of its energy. Consider for simplicity the conservative case when $\gamma = 0$. Since the energy is conserved, equation [4.11] can be written as:

$$\frac{m\dot{x}^2}{2} + W_p = W_0 \qquad [5.3]$$

where W_0 is the full or initial energy of the resonator. The velocity of the resonator can be expressed from this equation as follows:

$$\dot{x} = \pm\sqrt{\frac{2(W_0 - W_p)}{m}} \qquad [5.4]$$

By separating the variables due to x and t, we obtain $dt = dx/\sqrt{2m^{-1}(W_0 - W_p)}$. Let us assume that the turnoff points x_1 and x_2 bound the displacement of the system (see Figure 5.3). We can find the time required for the mass m to make one full oscillation between x_1 and x_2 and back:

$$T = 2\int_{x_1}^{x_2} \frac{dx}{\sqrt{2m^{-1}(W_0 - W_p)}} \qquad [5.5]$$

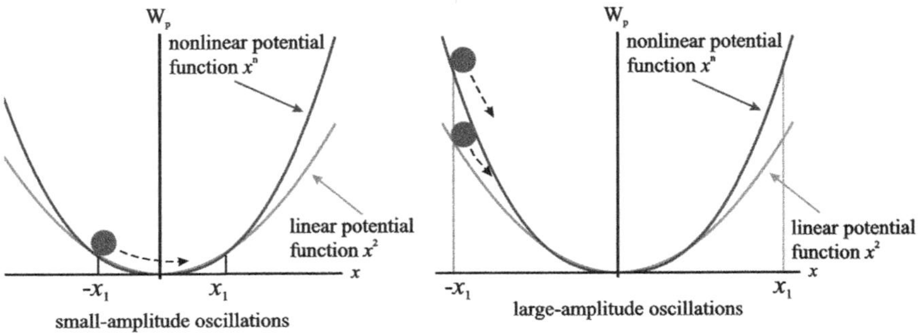

Figure 5.3. *Comparison of linear and nonlinear resonators using the analogy of a particle moving in the potential function W_p. In the case of small-amplitude oscillations when their amplitude x_1 is small, $W_{p,lin} \approx W_{p,nonlin}$ and the oscillations in the linear and nonlinear systems are similar. We would notice a difference for large-scale oscillations, when the two potential functions become different*

The quantity T defines the period of oscillations of a resonator whose potential energy is given by the function W_p and whose full energy is W_0. Fo symmetric potential functions W_p, which account for most cases, $x_2 = -x_1 = A_0$ where A_0 is the amplitude of oscillations.

It is possible to solve the integral in equation [5.5] for a linear resonator with $W_p = kx^2/2$. The time of one full oscillation in a linear resonator is constant $T = 2\pi/\sqrt{k/m} = 2\pi/\omega_0$ where, as usual, ω_0 is the natural frequency of the linear resonator. In the majority of other practical cases, including the Duffing resonator, the result of equation [5.5] will depend on the full energy W_0 of the resonator (or on the amplitude of oscillations A_0). For instance, for the Duffing resonator we can derive an approximate formula from equation [5.5]:

$$\omega_{nonlin} = \frac{2\pi}{T} \approx \sqrt{\frac{k}{m}}\left[1 + \frac{3(k_3/k_1)A_0^2}{8}\right] \qquad [5.6]$$

As seen from this expression, the larger the amplitude of oscillations A_0, the larger the frequency ω_{nonlin}. Next, we will study the dynamics of nonlinear resonators in time.

5.2. Semi-analytical techniques for nonlinear resonators

We will start this section by reminding the reader that there is no a general closed-form solution to a nonlinear equation. In particular, equations that describe nonlinear resonators, such as the Duffing resonator, do not have closed-form solutions. However, since the nonlinear Duffing resonator was introduced [DUF 18], a range of semi-analytical and perturbation methods that particularly suit the analysis of nonlinear resonators have been developed [NAY 93, NAY 08b, KUZ 02]. In particular, the most commonly used methods are as follows:

– straightforward expansion (SE);

– Lindstedt–Poincaré method (LPM);

– harmonic balance method (HBM);

– multiple scales method (MSM);

– averaging and the Van der Pol method.

Semi-analytical methods are not computationally intensive and allow us to analyze the behavior of a nonlinear system, both very quickly and accurately.

5.2.1. *Normalized form of nonlinear resonators*

In mathematics and physics, it is usual practice to normalize equations, i.e. to introduce new, often fully dimensionless, parameters and variables and rewrite the original equations using the new variables. There are two main reasons to apply normalization. First, realistic systems have many controlling and input parameters, and it is impossible to investigate the system behavior with respect to all of them. Thus, we are strongly motivated to reduce the number of the system parameters in order to ease the analysis. Moreover, in many cases we do not need to investigate the system with respect to each of the controlling parameters separately. Often, the behavior of a system depends on a combination or a ratio of two or more parameters, and the normalization procedure allows us to establish these important and meaningful combinations.

As an example, recall that a free linear resonator has three parameters: m, μ and k. However, all possible types of the resonator output are determined by only one dimensionless combination of parameters $\mu/(2\sqrt{k\,m})$. Below we are going to illustrate this statement. Taking the equation of the linear resonator

$$m\ddot{x} + \mu\dot{x} + kx = 0 \qquad [5.7]$$

we introduce the dimensionless time $\tau = \omega_0 t$ where $\omega_0 = \sqrt{k/m}$ (the natural frequency). Recall that $\dot{x} = \omega_0 x'$ and $\ddot{x} = \omega_0^2 x''$ where the prime now denotes the differentiation with respect to dimensionless time τ. The equation of the linear resonator can be transformed into an equation with only one parameter

$$x'' + 2\beta x' + x = 0 \qquad [5.8]$$

where $\beta = \mu/(2\sqrt{k\,m})$ plays the role of dimensionless damping. The natural frequency ω_0 has "disappeared" from the equation: it is now equal to unity. This means that we have chosen the new scale of time, which is related to the natural period of oscillations of the resonator. The normalization procedure shows that the dynamics of the resonator depend on the combination of its natural frequency, mass and damping (i.e. the dimensionless combination β) rather than on each three parameters independently.

If we have more terms in the equation of the linear resonator (for example an external force on the righthand side of the equation), we can apply another normalization. We can also normalize x by introducing a rescaled displacement $y = x/A$. Thus, starting from this moment, we will use the prime symbol and y in our equations to highlight that the equation has been somehow normalized:

$$y'' + 2\beta y' + y = \frac{F(t)}{m\omega_0^2\, A} \qquad [5.9]$$

In particular cases, we will be able to eliminate the coefficient that appears in the right-hand side of the equation. For instance, consider the case $F(t) = mA_{ext}\cos(\omega_{ext}t)$. We can choose $A = A_{ext}/\omega_0^2$ and eliminate the multiplicative coefficient before $\cos[(\omega_{ext}/\omega_0)\,\tau]$:

$$y'' + 2\beta y' + y = \cos(\Omega\tau) \qquad [5.10]$$

In this equation, we also denoted ω_{ext}/ω_0 as a normalized external frequency Ω.

Then, semi-analytical and perturbation methods are applied only if there is a "small" parameter that is present in a nonlinear equation. This means that we should have a very clear understanding on how to compare parameters. Many parameters cannot be compared directly. For example, how would we compare the mass m and the spring constant k? They describe completely different physical quantities. However, if we write a dimensionless combination of parameters, we can compare its "smallness" with respect to unity. Recall the dimensionless combination β we introduced above. It can be directly compared to unity. The linear resonator will have one type of behavior for $\beta < 1/2$ and another type of behavior for $\beta > 1/2$. We already know these two cases: ovedamped and underdamped oscillations.

Now we write the equation for the normalized Duffing resonator:

$$y'' + 2\beta y' + y + \lambda y^3 = 0 \qquad [5.11]$$

where we introduced $t = \omega_0 \tau$ and eliminated the coefficient ω_0. Note that as a result of normalization, we obtained the coefficient λ in front of the nonlinear term y^3. It is proportional to the parameter κ we used in Chapter 4 on the mechanical part of KEHs: $\lambda = \kappa A^2/\omega_0^2$.

Let us assume that β and λ are reasonably small ($\beta, \lambda \ll 1$). When we want to introduce a small parameter to be used with a semi-analytical technique, we simply present the parameters in the form: $\beta = \varepsilon\tilde{\beta}$ and $\lambda = \varepsilon\tilde{\lambda}$ where ε is an auxiliary small parameter, which we do not have to specify. This does not have a physical meaning, but rather serves for our convenience. Every time when we see ε in the equation, it will tell us that this term must be regarded as small. Thus, we obtain:

$$y'' + 2\varepsilon\tilde{\beta} y' + y + \varepsilon\tilde{\lambda} y^3 = 0 \qquad [5.12]$$

Now we proceed to selected methods of the analysis of nonlinear resonators. We should clearly state here that most of semi-analytical methods are applied to *weakly nonlinear systems*, i.e. systems with small nonlinear terms (in normalized equations). Although this does seem to be a serious limitation, many practical and realistic systems can be seen as weakly nonlinear, including nonlinear resonators for KEHs.

5.2.2. *Anharmonic oscillations demonstrated by straightforward expansion*

We are going to use the simplest approach, called the straightforward expansion, to show the first fundamental effect of nonlinearity – anharmonic oscillations. We will take the normalized Duffing resonator [5.12] where, for the simplicity of our analysis, we will assume no dissipation $\beta = 0$. Thus, we analyze the equation:

$$y'' + y + \varepsilon\tilde{\lambda} y^3 = 0 \qquad [5.13]$$

The SE method, like other semi-analytical methods we will discuss in this chapter, is based on presenting the solution to a nonlinear equation in the form of the following expansion:

$$y = y_0 + \varepsilon y_1 + \varepsilon^2 y_2 + \ldots \quad [5.14]$$

Let us substitute this into the original equation [5.13]:

$$(y_0 + \varepsilon y_1 + \varepsilon^2 y_2)'' + (y_0 + \varepsilon y_1 + \varepsilon^2 y_2) + \varepsilon\tilde{\lambda}(y_0 + \varepsilon y_1 + \varepsilon^2 y_2)^3 = 0 \quad [5.15]$$

where we truncate the series at ε^2 for the simplicity of analysis.

If we want the left-hand side of this equation to be equal to zero, all terms appearing before ε^0, ε^1, ε^2 must be equal to zero (since we deal with linearly independent combinations). And so we assume that

$$\begin{array}{ll} \varepsilon^0: & y_0'' + y_0 = 0 \\ \varepsilon^1: & y_1'' + y_1 = -\tilde{\lambda} y_0^3 \\ \varepsilon^2: & y_2'' + y_2 = -3\tilde{\lambda} y_0 y_1^2 \\ \ldots & \end{array} \quad [5.16]$$

We now have a set of linear "chain" equations with a non-zero right-hand side (i.e. inhomogeneous equations). By resolving the first equation, which is the equation of the linear resonator with the known solution, we can find y_0. We use it to resolve the second equation for y_1. Again, by finding y_1 and substituting it to the next equation, we find y_2, etc. Formally, we will have an infinite number of equations, but quite often it is enough to stop the expansion at some reasonable order.

It is also interesting to note that we start with the equation of the linear resonator. This means that we assume that the resulting nonlinear oscillations have a fundamental harmonic component and nonlinearities present in the equation distort in some way this fundamental component.

Let us undertake the first two steps of the method. The first equation in [5.16] is the linear resonator, and we know its solution (one of its forms):

$$y_0 = a\cos(\tau + \varphi) \quad [5.17]$$

where a and φ are some amplitude and phase of oscillations. We will remind you here that we are now using dimensionless time τ and dimensionless natural frequency $\omega_0 = 1$. Substituting y_0 into the second equation in set [5.16] we obtain:

$$y_1'' + y_1 = -\tilde{\lambda}a^3\cos^3(\tau + \varphi) = -\frac{\tilde{\lambda}a^3}{4}(3\cos(\tau + \varphi) + \cos(3\tau + 3\varphi)) \quad [5.18]$$

This is the equation of the linear resonator driven by two harmonic forces at frequencies $\omega_{ext} = 1$ and $\omega_{ext} = 3$. We can use solutions [3.51] and [3.52] to find the response. In order to use it, we set $F_0 = \tilde{\lambda} a^3/4$, $\omega_{ext} = 3$, $\gamma = 0$ and $\phi = -\pi$ and so the response to the $\cos(3\tau + 3\varphi)$ term is:

$$y_1^{(1)} = \frac{\tilde{\lambda}\, a^3}{32} \cos(3\tau + 3\varphi) \qquad [5.19]$$

The term $\cos(\tau + \varphi)$ represents the external force at the exact resonance frequency (i.e. the natural frequency $\omega_0 = 1$ of the linear undamped resonator on the left-hand side). This is the case of exact linear undamped resonance. The response to this term is:

$$y_1^{(2)} = -\frac{3\tilde{\lambda}\, a^3 \tau}{8} \sin(\tau + \varphi) \qquad [5.20]$$

and it is unlimitedly increasing with time τ. This term is called the linear resonance term, or secular term.

The overall solution, according to the superposition principle, is the sum of the individual responses:

$$y = y_0 + \varepsilon y_1 = y_0 + \varepsilon y_1^{(1)} + \varepsilon y_1^{(2)} \qquad [5.21]$$

and so

$$y(\tau) = a\cos(\tau + \varphi) + \frac{\lambda\, a^3}{32} \cos(3\tau + 3\varphi) - \frac{3\lambda\, a^3 \tau}{8} \sin(\tau + \varphi) \qquad [5.22]$$

In this equation, we returned to the original parameter λ again since $\varepsilon\tilde{\lambda} \equiv \lambda$. This shows that the solution depends only on the nonlinear parameter λ of the original equation, and the auxiliary parameter ε we introduced for convenience is eliminated from the final result.

From equation [5.22], we can see that the cubic nonlinearity of the original equation [5.13] has given rise to the third harmonic $\cos(3\tau + 3\varphi)$. This will result in an oscillation component at $3\omega_0$ when we return to our original variables and parameters. We can think of nonlinearity as a block in the system diagram that takes the input harmonic signal $x(t) = a_0 \cos(\omega_0 t)$ and transforms it into $\mathcal{F}(x(t))$. In particular, the most common type of nonlinearity, polynomial $\mathcal{F}(x(t)) = x^n$, will generate a series of harmonics at the output: 0 (DC), $2\omega_0$, $4\omega_0$, ..., $n\omega_0$ if n is even, and ω_0, $3\omega_0$, $5\omega_0$,..., $n\omega_0$ if n is odd. This example is shown in Figure 5.4.

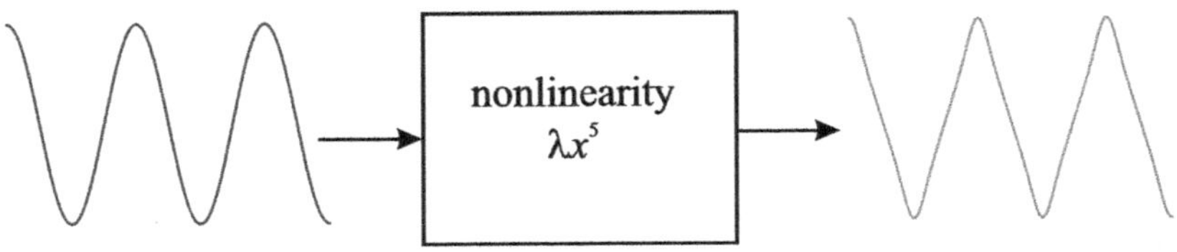

Figure 5.4. *Nonlinearity can be seen as a block in the system diagram. It takes a harmonic input $x(t) = a_0 \cos(\omega_0 t)$ and transforms it into the output $\mathcal{F}(x)$, generating a variety of different harmonics*

Despite its simplicity, the straightforward expansion is rarely applied. The main reason is, as you can see from solution [5.22], it leads to an infinite and uncompensated growth of the solution with time due to the term $\tau \sin(\tau + \varphi)$. The evolution of the solution with time is shown in Figure 5.5. On the other hand, it gives a very clear idea about anharmonic oscillations appearing as a result of nonlinearity.

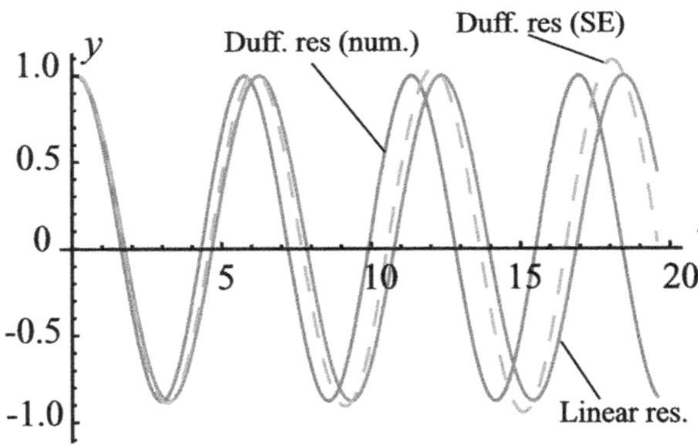

Figure 5.5. *Evolution of the normalized displacement y as a function of time for the linear resonator and the Duffing resonator* [5.13] *obtained from numerical simulation of the equation. For comparison, the approximated solution* [5.22] *obtained from the straightforward expansion is shown by the dashed line; $\omega_0 = 1$, $\lambda = 0.25$. For a color version of the figure, see www.iste.co.uk/basset/keh.zip*

5.2.3. *Anisochronous oscillations demonstrated by the LPM*

We know that in real systems oscillations do not increase infinitely. The initial growth of the displacement, velocity or any other physical quantity will be stopped by either air damping, dissipation or nonlinearity. Nonlinearity plays a limiting role in physical systems. In particular, in the context of nonlinear resonators, the mechanism through which the growth of the displacement is stopped is because the frequency of nonlinear oscillations depends on the energy of the resonator (i.e. nonlinear oscillations are anisochronous).

We have briefly described this effect in the previous section. In this section, we suggest a simple and effective way to modify the straightforward expansion method to take into account this effect. Moreover, we will obtain the approximation of the nonlinear frequency ω_{nonlin} of oscillations in the Duffing resonator.

The method is based on introducing one more time scale $\hat{t} = \omega\tau$. Note that this can be done even though we have already been using normalized dimensionless time τ and frequency $\omega_0 = 1$. The idea of the new time scale is to introduce a correction factor ω that will describe anisochronous oscillations. Recalling that $d/d\tau$ turns into $\omega^2 d/d\hat{t}$, we have:

$$\omega^2 y'' + y + \varepsilon\tilde{\lambda}y^3 = 0 \qquad [5.23]$$

where we still use the prime to denote the derivative with respect to new dimensionless time $\hat{t}$. We will present not only the displacement y, but also the new frequency ω as a series:

$$\begin{aligned} y &= y_0 + \varepsilon y_1 + \varepsilon^2 y_2 + \ldots \\ \omega &= 1 + \varepsilon\omega_1 + \varepsilon^2\omega_2 + \ldots \end{aligned} \qquad [5.24]$$

Because we already use a normalized time scale τ, so that the natural frequency of linear oscillations in the system is $\omega_0 = 1$, we present the new frequency ω as a power series around the unity frequency. One could apply this to a dimensional non-normalized resonator with the natural frequency $\omega_0 \neq 1$. In this case, we would expand the new frequency as $\omega = \omega_0 + \varepsilon\omega_1 + \varepsilon^2\omega_2 + \ldots$.

Substituting both series to the original equation [5.23], we obtain:

$$\begin{aligned} (1 + \varepsilon\omega_1 + \varepsilon^2\omega_2)^2(y_0'' + \varepsilon y_1'' + \varepsilon^2 y_2'') + (y_0 + \varepsilon y_1 + \varepsilon^2 y_2) \\ + \varepsilon\tilde{\lambda}(y_0 + \varepsilon y_1 + \varepsilon^2 y_2)^3 = 0 \end{aligned} \qquad [5.25]$$

where we truncate the series at ε^2 for the simplicity of analysis. Again, setting all terms of ε^n to zero, we obtain the system of equations:

$$\begin{aligned} y_0'' + y_0 &= 0 \\ y_1'' + y_1 &= -2\omega_1 y_0'' - \tilde{\lambda}y_0^3 \end{aligned} \qquad [5.26]$$

We already know that the solution to y_0 has the form:

$$y_0 = a\cos(\hat{t} + \varphi) \qquad [5.27]$$

Substituting this to the second equation in [5.26], we obtain:

$$y_1'' + y_1 = 2\omega_1 a\cos(\hat{t} + \varphi) - \frac{3\tilde{\lambda}\,a^3}{4}\cos(\hat{t} + \varphi) - \frac{\tilde{\lambda}\,a^3}{4}\cos(3\hat{t} + 3\varphi) \qquad [5.28]$$

The problem with the SE method was caused by the presence of the term that leads to linear undamped resonance (i.e. the terms proportional to $\cos(\tau+\varphi)$ in the equation for y_1). Since we have introduced a new time scale $\hat{t}$ using one new parameter ω, we can provide one additional condition on this parameter. Since in a real system linear undamped resonance is not possible, we add such a condition on ω so the resonance terms are eliminated from the equation:

$$2\omega_1 a\cos(\hat{t} + \varphi) - \frac{3\tilde{\lambda}\,a^3}{4}\cos(\hat{t} + \varphi) = 0 \qquad [5.29]$$

and so we find ω_1:

$$\omega_1 = \frac{3\tilde{\lambda}\,a^2}{8} \qquad [5.30]$$

with the resonance terms removed, we solve the equation:

$$y_1'' + y_1 = -\frac{\tilde{\lambda}\,a^3}{4}\cos(3\tau + 3\varphi) \qquad [5.31]$$

The particular solution of this equation that correspond to a forced oscillation under a harmonic force $F(t) = (\tilde{\lambda}a^3/4)\cos(3\tau + 3\varphi)$ has been found in section 5.2.2:

$$y_1 = \frac{\tilde{\lambda}\,a^3}{32}\cos(3\hat{t} + 3\varphi) \qquad [5.32]$$

Thus, the overall solution is

$$\begin{aligned} y(\tau) \approx y_0 + \varepsilon y_1 &= a\cos(\hat{t} + \varphi) + \tfrac{\lambda a^3}{32}\cos(3\hat{t} + 3\varphi) \\ &= a\cos(\omega\tau + \varphi) + \tfrac{\lambda a^3}{32}\cos(3\omega\tau + 3\varphi) \end{aligned} \qquad [5.33]$$

where we have returned to time τ and explicitly used the correction factor for the frequency:

$$\omega_{nonlin} \equiv \omega \approx 1 + \frac{3\lambda a^2}{8} \qquad [5.34]$$

Note again that the resulting solution is independent of the small parameter ε as it should be.

Clearly, with this approach we have taken into account both, anharmonic oscillations (assuming the power series for y) and anisochronous oscillations (by introducing the corrected frequency ω). Note that the corrected frequency [5.34] has the same form as we obtained from [5.6]. The LPM is accurate enough to capture these nonlinear phenomena. The comparison of the semi-analytical solution [5.33] with the numerical solution of the Duffing equation [5.13] is shown in Figure 5.6.

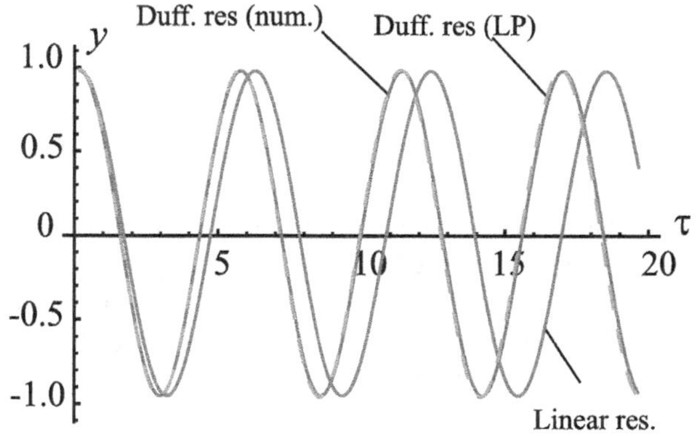

Figure 5.6. *Evolution of the normalized displacement y as a function of time for the linear resonator and the Duffing resonator [5.13] obtained from numerical simulation of the equation. For the comparison, approximated solution [5.33] obtained from the LPM is shown by the dashed line; $\omega_0 = 1$, $\lambda = 0.25$. For a color version of the figure, see www.iste.co.uk/basset/keh.zip*

5.2.4. *Multiple scales method*

There are a range of advanced methods that could be used for resonators with a more general form of nonlinearity or with parametric actuation. We are going to consider one of them in this chapter – the multiple scale method (MSM). This is the most advanced method compared to the straightforward expansion and the LPM. The MSM also takes into account anharmonic and anisochronous oscillations. While the LPM corrects the frequency of nonlinear oscillations by assuming a parameter ω (the corrected frequency), the MSM introduces *time scales*. The idea is somewhat similar: we will try to take into account the change in the frequency of nonlinear oscillations. In addition, if you examine the result of the SE and LPM, you will see that we found the steady-state solution to a nonlinear resonator. Unfortunately, we were not able to obtain the information about the transient process with these methods. Another powerful method, the harmonic balance method (HBM) that we mentioned at the beginning of this chapter also does not provide this information. The strong advantage of the MSM is that we will be able to obtain the form of the transient process in a nonlinear resonator.

We will analyze the Duffing equation using the MSM, but this time we will include a dissipation term in our analysis:

$$y'' + 2\varepsilon\tilde{\beta}y' + y + \varepsilon\tilde{\lambda}y^3 = 0 \qquad [5.35]$$

The method is based on introducing a number of time scales that are denoted as:

$$T_0 = t, \quad T_1 = \varepsilon t, \quad T_2 = \varepsilon^2 t... \qquad [5.36]$$

Indeed, the solution to [5.35] contains components that evolve differently with time. There is a component that is related to eigen (free) oscillations of a weakly nonlinear resonator. The characteristic time of this component is $2\pi/\omega_0 = 2\pi$ (since now $\omega_0 = 1$) and it corresponds to the fundamental time scale T_0. Another component is related to the decay of oscillations due to the small dissipation term $2\varepsilon\tilde{\beta}y'$. The characteristic time of that component is $1/(\varepsilon\tilde{\beta})$, much slower than the fundamental time scale T_0, and it corresponds to the first slow time scale T_1.

Thus, the original variable $y(\tau)$ of a single variable τ in [5.35] turns into a variable of multiple times T_i. As usual, we present it as a series, and every term in this series is a function of the newly introduced time scales:

$$y(\tau) = y_0(T_0, T_1, T_2, ...) + \varepsilon y_1(T_0, T_1, T_2, ...) + \varepsilon^2 y_2(T_0, T_1, T_2, ...) + ... \qquad [5.37]$$

In order to use this in our original differential equation, we must find the full derivative with respect to time τ. We use the formula that expresses a full derivative through partial derivatives with respect to times T_i:

$$\begin{aligned}\frac{d}{d\tau} &= \frac{\partial}{\partial T_0}\frac{dT_0}{d\tau} + \frac{\partial}{\partial T_1}\frac{dT_1}{d\tau} + \frac{\partial}{\partial T_2}\frac{dT_2}{d\tau} + ... \\ &= \frac{\partial}{\partial T_0} + \varepsilon\frac{\partial}{\partial T_1} + \varepsilon^2\frac{\partial}{\partial T_2} + ... \\ &= D_0 + \varepsilon D_1 + \varepsilon^2 D_2 ... \\ \frac{d^2}{d\tau^2} &= (D_0 + \varepsilon D_1 + \varepsilon^2 D_2 ...)^2 \\ &= D_0^2 + 2\varepsilon D_0 D_1 + \varepsilon^2(D_1^2 + 2D_0 D_2) + ...\end{aligned} \quad [5.38]$$

where we introduced the operators $D_i = \partial/\partial T_i$. Using equations [5.37] and [5.38], we rewrite the original equation [5.35]:

$$\begin{aligned}&\left(D_0^2 + 2\varepsilon D_0 D_1 + \varepsilon^2(D_1^2 + 2D_0 D_2) + ...\right)\left(y_0 + \varepsilon y_1 + \varepsilon^2 y_2 + ...\right) \\ &+2\varepsilon\tilde{\beta}\left(D_0 + \varepsilon D_1 + \varepsilon^2 D_2 ...\right)\left(y_0 + \varepsilon y_1 + \varepsilon^2 y_2 + ...\right) \\ &+(y_0 + \varepsilon y_1 + \varepsilon^2 y_2 + ...) + \varepsilon\tilde{\lambda}(y_0 + \varepsilon y_1 + \varepsilon^2 y_2 + ...)^3 = 0\end{aligned} \quad [5.39]$$

As the next step, we set every ε^n order to zero (we will, for convenience, use only the first two terms of the expansion):

$$\begin{aligned}&D_0^2 y_0 + y_0 = 0 \\ &D_0^2 y_1 + y_1 = -2D_0 D_1 y_0 - 2\tilde{\beta} D_0 y_0 - \tilde{\lambda} y_0^3\end{aligned} \quad [5.40]$$

Now, we will introduce the complex form of variables, which is immensely useful when we deal with polynomial nonlinearities. From the first equation in [5.40], we obtain that:

$$y_0 = A(T_1)e^{jT_0} + A^*(T_1)e^{-jT_0} = a(T_1)\cos(\tau + \varphi(T_1)) \quad [5.41]$$

where $A(T_1) = (a/2)\exp(j\varphi)$ is the slow amplitude of oscillations, A^* is the complex conjugate of A and φ is the phase of oscillations. As you can see, we again obtain a harmonic oscillation as the first approximation to our solution. However, we will keep it in the complex form.

Now we will use it in the second equation in [5.40], taking into account that $D_0 y_0 = jA(T_1)\exp(jT_0) - jA^*(T_1)\exp(-jT_0)$ and $D_1 y_0 = \dot{A}(T_1)\exp(jT_0) +$

$\dot{A}^*(T_1)\exp(-jT_0)$, where $\dot{A}$ now denotes the derivative with respect to time scale T_1. Thus, we have:

$$\begin{aligned}D_0^2 y_1 + y_1 &= -2j\dot{A}e^{jT_0} + 2j\dot{A}^*e^{-jT_0} - 2j\tilde{\beta}Ae^{jT_0} + 2j\tilde{\beta}A^*e^{-jT_0} \\ &-\tilde{\lambda}(Ae^{jT_0} + A^*e^{-jT_0})^3\end{aligned} \qquad [5.42]$$

Instead of one real-valued variable y, we introduced a complex variable A (that contains in fact two variables, $\mathrm{Re}(A)$ and $\mathrm{Im}(A)$ or a and φ). Therefore, we can suggest one additional equation for the variable A (similar to the LPM). In order to introduce such an equation, let us analyze the terms on the right-hand side of the above equation. The terms that are proportional to $\exp(jT_0)$ lead to resonance in the linear undamped system (as discussed previously in the LP method). These terms cause an infinite increase in the oscillation amplitude. We know that this does not happen in the real system. We will choose an additional equation that will eliminate these terms. Thus, we state that

$$-2j\dot{A}e^{jT_0} - 2j\tilde{\beta}Ae^{jT_0} - 3\tilde{\lambda}A^2A^*e^{jT_0} = 0 \qquad [5.43]$$

This equation allows us to find the slow amplitude A (note that equation [5.43] automatically implies that its complex conjugate expression is also zero). As a result, all terms proportional to e^{jT_0} and e^{-jT_0} are eliminated from equation [5.42].

Expressing complex A through real a and φ, from [5.43] we obtain two equations that define the evolution of slow amplitude a and phase φ with time:

$$\begin{aligned}\dot{a} &= -\tilde{\beta}a \\ \dot{\varphi} &= -3\tilde{\lambda}a^2/8\end{aligned} \qquad [5.44]$$

These equations are called the truncated equations. They allow us to find the evolution of the slow amplitude and phase with time, as well as their steady-state values. Knowing the initial conditions $a(0) = a_0$ and $\varphi(0) = \varphi_0$, we solve these equations:

$$\begin{aligned}a &= a_0 e^{-\tilde{\beta}T_1} \\ \dot{\varphi} &= \varphi_0 + \frac{3\tilde{\lambda}a_0^2}{16\tilde{\beta}}\left(1 - e^{-2\tilde{\beta}T_1}\right)\end{aligned} \qquad [5.45]$$

Finally, in order to find y, we return to equation [5.42] where we excluded all resonant terms:

$$D_0^2 y_1 + y_1 = -(\tilde{\lambda}a^3/4)\cos(3\tau + 3\varphi) \qquad [5.46]$$

In [5.46], we have taken into account that the only two terms left in the right-hand side are $A^3 \exp(3jT_0) + (A^*)^3 \exp(-3jT_0) = (a^3/4)\cos(3\tau + 3\varphi)$. We already found the particular solution of equation [5.46] in section 5.2.2:

$$y_1 = \frac{\tilde{\lambda} a^3}{32} \cos(3\tau + 3\varphi) \quad [5.47]$$

Now we write the overall solution:

$$\begin{aligned} y(\tau) &= y_0 + \varepsilon y_1 = a_0 \cos(\tau + \varphi) + \tfrac{\lambda a_0^3}{32} \cos(3\tau + 3\varphi) \\ &= a_0 e^{-\beta\tau} \cos\left(\tau + \varphi_0 + \frac{3\lambda a_0^2}{16\beta}\left(1 - e^{-2\beta\tau}\right)\right) \\ &\quad + \frac{\lambda a_0^3 e^{-3\beta\tau}}{32} \cos\left(3\tau + 3\varphi_0 + \frac{9\lambda a_0^2}{16\beta}\left(1 - e^{-2\beta\tau}\right)\right) \end{aligned} \quad [5.48]$$

where we again took into account that $\varepsilon\tilde{\beta} = \beta$, $\varepsilon\tilde{\lambda} = \lambda$ and $\tilde{\lambda}/\tilde{\beta} = \lambda/\beta$.

Let us look at this solution closely and compare it with solution [5.33] from the LPM. We see that equation [5.48] does not give us the frequency of nonlinear oscillations in the explicit form. However, we can check whether or not equation [5.48] turns into equation [5.33] in the conservative case $\beta \to 0$. Indeed, $1 - \exp(-2\beta\tau) \approx 2\beta\tau + O(t^2)$ and so we obtain

$$y(\tau) \approx a_0 \cos\left(\left(1 + \tfrac{3\lambda a^2}{8}\right)\tau + \varphi_0\right) + \tfrac{\lambda a_0^3}{32} \cos\left(3\left(1 + \tfrac{3\lambda a^2}{8}\right)\tau + 3\varphi_0\right) \quad [5.49]$$

The MSM provides an accurate approximation to the solution of the original Duffing equation as shown in Figure 5.7. The MSM is a very flexible and powerful method that can be adjusted to various types of nonlinearities and external forces. It also allows us to find the transient solution and analyze for stability.

5.2.5. *Nonlinearity of a general form*

So far we have used the Duffing resonator as an example of a nonlinear system. From the three methods, demonstrated in the earlier sections, the case of any polynomial nonlinearity y^n can be analyzed in a very similar manner. However, we may meet other types of nonlinear functions (such as the transducer force F_t) that can be, in addition to their nonlinear (and non-polynomial) dependence on the displacement y, piecewise defined. For this reason, let us briefly outline how to use the MSM in the case of a general nonlinear term (nonlinear function) $G(y)$. We will consider the equation:

$$y'' + 2\varepsilon\tilde{\beta}y' + y + \varepsilon\tilde{\lambda}G(y) = 0 \quad [5.50]$$

where again we highlight that both terms, $\tilde{\beta}y'$ and $\tilde{\lambda}G(y)$, appear to be small. We use the power series [5.37] and [5.38] in equation [5.50] to obtain:

$$\begin{aligned}&\left(D_0^2 + 2\varepsilon D_0 D_1 + \varepsilon^2(D_1^2 + 2D_0 D_2) + ...\right)\left(y_0 + \varepsilon y_1 + \varepsilon^2 y_2 + ...\right)\\ &+2\varepsilon\tilde{\beta}\left(D_0 + \varepsilon D_1 + \varepsilon^2 D_2 ...\right)\left(y_0 + \varepsilon y_1 + \varepsilon^2 y_2 + ...\right)\\ &+(y_0 + \varepsilon y_1 + \varepsilon^2 y_2 + ...) + \varepsilon\tilde{\lambda}G(y_0 + \varepsilon y_1 + \varepsilon^2 y_2 + ...) = 0\end{aligned} \quad [5.51]$$

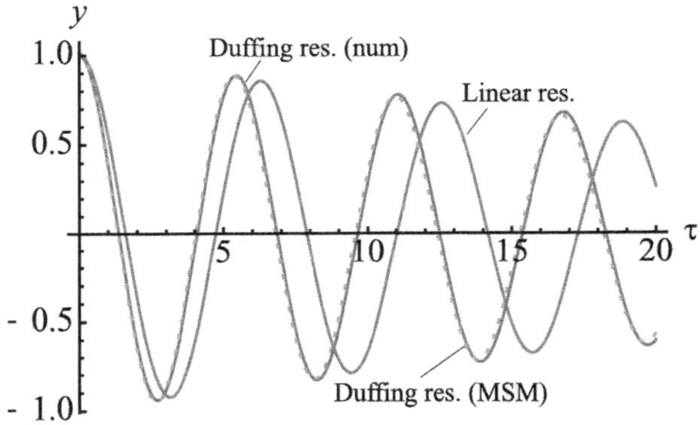

Figure 5.7. *Evolution of the normalized displacement y as a function of time for the linear resonator and the Duffing resonator* [5.13] *obtained from numerical simulation of the equation. For the comparison, approximated solution* [5.48] *obtained from the MSM is shown by the dashed line; $\omega_0 = 1$, $\lambda = 0.25$. For a color version of the figure, see www.iste.co.uk/basset/keh.zip*

We represent the nonlinear term $G(y)$ using the Taylor series:

$$\varepsilon\,\tilde{\lambda}\,G(y_0 + \varepsilon y_1 + ...) \approx \varepsilon\,\tilde{\lambda}\,G(y_0) + \varepsilon^2\,\tilde{\lambda}\,G'(y_0)y_1 + ... \quad [5.52]$$

where $G'(y_0)$ is the first derivative of $G(y)$ with respect to y.

Again, collecting the terms ε^n, we obtain a set of chain equations (only the first two equations are written below for convenience):

$$\begin{aligned}&D_0^2 y_0 + y_0 = 0\\ &D_0^2 y_1 + y_1 = -2D_0 D_1 y_0 - 2\tilde{\beta}D_0 y_0 - \tilde{\lambda}G(y_0)\end{aligned} \quad [5.53]$$

Note that the first equation does not change and will result in a harmonic solution:

$$y_0 = A(T_1)e^{jT_0} + A^*(T_1)e^{-jT_0} = a(T_1)\cos(\tau + \varphi(T_1)) \qquad [5.54]$$

where $A(T_1) = (a/2)\exp(j\varphi)$ is the slow amplitude of oscillations, A^* is the complex conjugate of A and φ is the phase of oscillations. Until now, the steps that we undertake are very similar to the case of the Duffing resonator.

At this stage, it is useful to note that since y_0 is a harmonic (periodic) function of time, the nonlinear function $G(y_0)$ will also be periodic. If so, we can present it using the truncated Fourier series:

$$G(y_0) \approx G_0 + \sum_{n=1}^{N} a_n(y_0)\cos n(\tau + \varphi) + \sum_{n=1}^{N} b_n(y_0)\sin n(\tau + \varphi) \qquad [5.55]$$

where

$$\begin{aligned} G_0 &= \tfrac{1}{2\pi}\int_0^{2\pi} (a\cos\theta)d\theta \\ a_n &= \tfrac{1}{\pi}\int_0^{2\pi} (a\cos\theta)\cos n\theta d\theta \\ b_n &= \tfrac{1}{\pi}\int_0^{2\pi} (a\cos\theta)\sin n\theta d\theta \end{aligned} \qquad [5.56]$$

We write the second equation in set [5.53] in the following form:

$$\begin{aligned} D_0^2 y_1 + y_1 = &-2D_0D_1y_0 - 2\tilde{\beta}D_0y_0 - \tilde{\lambda}G_0 \\ &-\tilde{\lambda}\sum_{n=1}^{N} a_n(y_0)\cos n(\tau + \varphi) - \tilde{\lambda}\sum_{n=1}^{N} b_n(y_0)\sin n(\tau + \varphi) \end{aligned} \qquad [5.57]$$

Thus, the general nonlinear function $G(y)$ serves as an external, periodic excitation force that drives a linear lossless resonator with the displacement y_1. The coefficients a_n and b_n, which are the amplitudes of the cosine and sine components of $G(y)$, depend on the amplitude of the resonators through $y_0 = a\cos(\tau + \varphi)$. This is an expected result since the function $G(y)$ depends on the displacement y nonlinearly.

In addition, due to the Fourier representation [5.55], we can see how nonlinearity gives rise to multiple harmonics of the fundamental frequency at the output of the nonlinear resonator. In practice, however we can restrict ourselves and consider only the fundamental harmonic in the Fourier series [5.55]. We know that a linear high-Q resonator does not respond effectively to higher harmonics of a driving external force. It is not difficult for us to find the response to every harmonic on the right-hand side of equation [5.57] (as discussed in Chapter 3). However, this response will be significantly smaller than the response of the resonator at the fundamental resonance. The qualitative explanation is given in Figure 5.8 where we show that the linear high-Q resonator acts as a very selective filter. Thus, without the loss of generality, we can limit the series [5.55] to the fundamental harmonic:

$$\begin{aligned} G(y_0) &= G_0 + a_1(y_0)\cos(\tau+\varphi) + b_1(y_0)\sin(\tau+\varphi) \\ &= G_0 + c_1(y_0)e^{jT_0+j\varphi} + c.c. \end{aligned} \quad [5.58]$$

where

$$c_1 = \frac{a_1 - ib_1}{2} \quad [5.59]$$

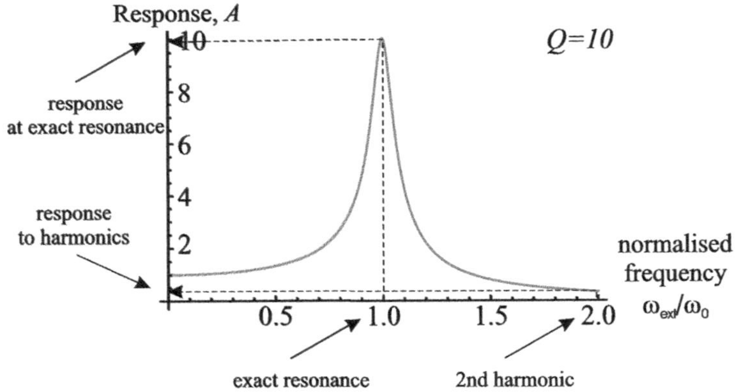

Figure 5.8. *The concept of a high-Q resonator (by the example of the resonator with $Q = 10$). Due to its high selectivity, the linear resonator with reasonably large quality factors responds effectively only to the frequencies that lie in the resonance band, i.e. in the vicinity of the natural frequency of the resonator ω_0*

In equation [5.58], $c.c.$ stands for complex conjugate. Using approach [5.58], we can analyze the effect of nonlinear forces of a general form (not necessary polynomial). The corresponding term that contains c_1 will appear in [5.53]. We can study the resulting equation in the same manner as is described in section 5.2.4.

In order to validate this approach, let us treat the Duffing type nonlinearity of the form $G(y) = y^3$ using the Fourier series. As a result, we should obtain the result of the earlier section. According to the second equation in set [5.53], we are to find the Fourier coefficients of $G(y_0) = (a\cos(\tau + \varphi))^3$, and so the representation [5.55] yields the following series for this force:

$$\begin{aligned} G(y_0) &= \tfrac{3a^3}{4}\cos(\tau + \varphi) + \tfrac{a^3}{4}\cos(3\tau + 3\varphi) \\ &= \left[\tfrac{3a^3}{8}e^{j(T_0+\varphi)} + \tfrac{a^3}{8}e^{3j(T_0+\varphi)} + c.c.\right] \end{aligned} \qquad [5.60]$$

substituting this into [5.53] and setting the secular terms to zero we obtain:

$$-2j\dot{A}e^{jT_0} - 2j\tilde{\beta}Ae^{jT_0} - \frac{3\tilde{\lambda}a^3}{8}e^{j(T_0+\varphi)} = 0 \qquad [5.61]$$

We again express the complex amplitude A through real a and φ (as usual, $A = (a/2)\exp[j\varphi]$) and separate the real and imaginary part of this equation. We obtain two truncated equations on a and φ that are identical to [5.44]. Thus, we received the same result, and the approach that uses the Fourier series is equivalent to the standard implementation of the MSM given in section 5.2.4.

5.3. Concluding remarks

Thus, using selected semi-analytical techniques, we have shown the fundamental effects of nonlinearity – anharmonic and anisochronous oscillations. The straightforward expansion shows how a particular type of nonlinearity gives rise to specific harmonics in the response of a nonlinear resonator. The LPM demonstrates how to obtain an explicit correction for the frequency associated with nonlinear oscillations. Finally, the multiple scale method is a very flexible tool that allows us to analyze a broad range of nonlinear resonators. In the next chapter, we will apply this knowledge to analyze nonlinear resonance and discuss its application to electrostatic kinetic energy harvesters.

6

Nonlinear Resonance and its Application to Electrostatic Kinetic Energy Harvesters

In this chapter, we discuss driven nonlinear resonators and nonlinear resonance in detail. We derive the equation of the nonlinear resonance curve and explain how to extend our analysis to the case of a general form of nonlinearity. We show how electrostatic kinetic energy harvesters can be analyzed from the stand point of nonlinear resonators and resonance, and how nonlinear resonance gives rise to electrostatic softening and hardening effects.

6.1. Forced nonlinear resonator and nonlinear resonance

6.1.1. *Analysis of forced oscillations using the multiple scales method*

We have already studied a linear resonator under external harmonic actuation and established that if the frequency of the external force is close to the natural frequency of the resonator, one observes resonance. In a nonlinear resonator, as we know from the discussion in Chapters 4 and 5, oscillations are anisochronous, i.e. the frequency of such oscillation is not a constant, but rather depends on the energy of the resonator. However, resonance is also possible in nonlinear systems. The resulting forced oscillations are more complex than in the linear case. We will consider the Duffing resonator as an example to illustrate nonlinear resonance. We will begin by writing a forced Duffing resonator in the following dimensionless form:

$$y'' + 2\beta y' + y + \lambda y^3 = \alpha \cos((1+\sigma)\tau + \phi_0) \qquad [6.1]$$

We remind the readers that λ can be positive or negative. The term $\cos[(1 + \sigma)\tau + \phi_0]$ appeared after the normalization procedure. Indeed,

$\cos(\omega_{ext}t + \phi_0) = \cos(\omega_{ext}(\omega_0^{-1}\tau) + \phi_0) = \cos(\Omega\tau + \phi_0)$ with $\Omega = \omega_{ext}/\omega_0$. We will present the dimensionless external frequency in the form $\Omega = 1 + \sigma$, where $\sigma = \omega_{ext}/\omega_0 - 1$ is a frequency mismatch between the dimensionless natural frequency (equal to unity after normalization) and the dimensionless external frequency is Ω. As a result of normalization, we introduce the dimensionless acceleration of the external force $\alpha = A_{ext}/(\omega_0^2 A)$. As in the previous chapter, following the procedure of introducing a small parameter ε, we will assume that $\beta = \varepsilon\tilde{\beta}$, $\alpha = \varepsilon\tilde{\lambda}$, $\sigma = \varepsilon\tilde{\sigma}$ and $\alpha = \varepsilon\tilde{\alpha}$. This results in the following equation:

$$y'' + 2\varepsilon\tilde{\beta}y' + y + \varepsilon\tilde{\lambda}y^3 = \varepsilon\tilde{\alpha}\cos((1 + \varepsilon\tilde{\sigma})\tau + \phi_0) \qquad [6.2]$$

We will apply the multiple scales method (MSM) following the procedure described in section 5.2.4, limiting our power series to the first and the second terms. The introduction of time scales leads to the equation:

$$\begin{aligned} &\left(D_0^2 + 2\varepsilon D_0 D_1\right)(y_0 + \varepsilon y_1) \\ &+2\varepsilon\tilde{\beta}\left(D_0 + \varepsilon D_1\right)(y_0 + \varepsilon y_1) + (y_0 + \varepsilon y_1) \\ &+\varepsilon\tilde{\lambda}(y_0 + \varepsilon y_1)^3 = \varepsilon\tilde{\alpha}\cos((1 + \varepsilon\tilde{\sigma})\tau + \phi_0) \end{aligned} \qquad [6.3]$$

The term of the external force can also be presented using the complex form: $\tilde{\alpha}\cos((1 + \varepsilon\tilde{\sigma})\tau + \phi_0) = (\tilde{\alpha}/2)\exp[j(1 + \varepsilon\tilde{\sigma})\tau + j\phi_0] + c.c.$

Collecting the terms corresponding to different orders of ε, we obtain:

$$\begin{aligned} &D_0^2 y_0 + y_0 = 0 \\ &D_0^2 y_1 + y_1 = -2D_0 D_1 y_0 - 2\tilde{\beta}D_0 y_0 - \tilde{\lambda}y_0^3 \\ &\qquad\qquad + ((\tilde{\alpha}/2)\exp[j(1 + \varepsilon\tilde{\sigma})T_0 + j\phi_0] + c.c.) \end{aligned} \qquad [6.4]$$

The solution to the first equation of the set [6.4] has the form $y_0 = A(T_1)\exp(jT_0) + c.c.$ where the complex slow amplitude $A = (a/2)\exp(j\varphi)$ can be written through a real slow amplitude a and real slow phase φ. In the second equation of the set [6.4], we set all secular terms to zero in order to avoid a linear resonance in a lossless system. This yields the equation for A:

$$-2j\dot{A}e^{jT_0} - 2j\tilde{\beta}Ae^{jT_0} - 3\tilde{\lambda}A^2A^*e^{jT_0} + (\tilde{\alpha}/2)e^{j\varepsilon\tilde{\sigma}T_0 + j\phi_0}e^{jT_0} + c.c = 0 \qquad [6.5]$$

Note that the term $\exp(j\varepsilon\tilde{\sigma}T_0 + j\phi_0) = \exp(j\tilde{\sigma}T_1 + j\phi_0)$ is in fact a slowly varying term, as $T_1 = \varepsilon T_0$. Now using the representation of the complex A through real a and φ, we obtain the equations:

$$\begin{aligned} \dot{a} &= -\tilde{\beta}a + \tfrac{\tilde{\alpha}}{2}\sin(\tilde{\sigma}T_1 + \phi_0 - \varphi) \\ a\dot{\varphi} &= \tfrac{3\tilde{\lambda}a^3}{8} - \tfrac{\tilde{\alpha}}{2}\cos(\tilde{\sigma}T_1 + \phi_0 - \varphi) \end{aligned} \qquad [6.6]$$

where we remind the readers that the overdot denotes the derivative with respect to the slow time scale T_1. Finally, we will introduce a new variable phase $\psi = \tilde{\sigma}T_1 + \phi_0 - \varphi$. Since $\dot{\psi} = \tilde{\sigma} - \dot{\phi}$, equations [6.6] will take the form:

$$\begin{aligned} \dot{a} &= -\tilde{\beta}a + \tfrac{\tilde{\alpha}}{2}\sin\psi \\ \dot{\psi} &= \tilde{\sigma} - \tfrac{3\tilde{\lambda}a^2}{8} + \tfrac{\tilde{\alpha}}{2a}\cos\psi \end{aligned} \qquad [6.7]$$

We can compare the set of equations [6.7] to the truncated equations [5.44] of a free resonator. The external driving is presented by the terms proportional to $\tilde{\alpha}$. As a result, forced oscillations do not decay to zero. After a transient process, oscillations in a nonlinear resonator reach steady-state with a constant amplitude and phase. Thus, from the above equations, we will be able to find the amplitude a_0 and phase ψ_0 of steady-state forced oscillations by assuming that $a_0, \psi_0 = \text{const}$, i.e. $\dot{a_0}, \dot{\psi_0} = 0$:

$$\begin{aligned} \tfrac{\tilde{\alpha}}{2}\sin\psi_0 &= \tilde{\beta}a_0 \\ \tfrac{\tilde{\alpha}}{2a_0}\cos\psi_0 &= -\tilde{\sigma} + \tfrac{3\tilde{\lambda}a_0^2}{8} \end{aligned} \qquad [6.8]$$

Since $\cos^2\psi + \sin^2\psi = 1$, by squaring both equations in [6.8] and adding them together, we obtain the equation that allows us to find the slow real amplitude a_0:

$$\beta^2 a_0^2 + \left(\sigma a_0 - \frac{3\lambda a_0^3}{8}\right)^2 = \frac{\alpha^2}{4} \qquad [6.9]$$

In this equation, we used that $\tilde{\beta} = \beta/\varepsilon$, $\tilde{\sigma} = \sigma/\varepsilon$, $\tilde{\lambda} = \alpha/\varepsilon$ and $\tilde{\alpha} = \alpha/\varepsilon$. We have already found that the solution to [6.4] is:

$$\begin{aligned} y(\tau) &= a_0\cos(\tau + \varphi_0) + \tfrac{\alpha a_0^3}{32}\cos(3(1+\sigma)\tau + 3\phi_0 - 3\psi_0) \\ &= a_0\cos(\Omega\tau + \phi_0 - \psi_0) + \tfrac{\lambda a_0^3}{32}\cos(3\Omega\tau + 3\phi_0 - 3\psi_0) \end{aligned} \qquad [6.10]$$

with a_0 and ψ_0 obtained from the steady-state equations [6.8] and [6.9]. The analysis of this expression shows that forced oscillations in a nonlinear resonators are at the frequency of the external vibrations Ω.

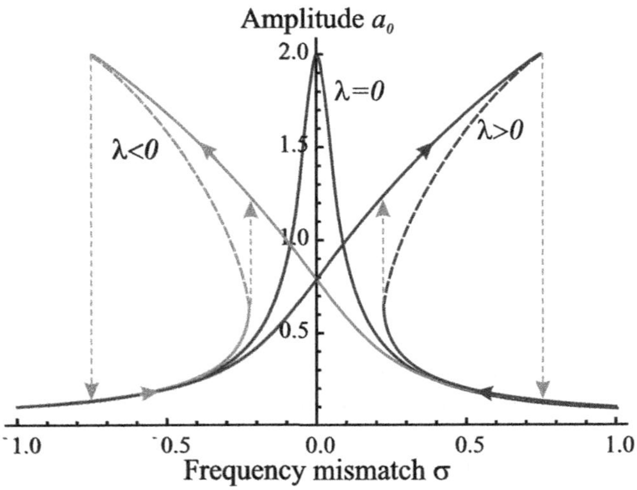

Figure 6.1. *Resonance in linear and nonlinear resonators. The amplitude a_0 of steady-state forced oscillations as a function of the normalized frequency mismatch $\sigma = \omega_{ext}/\omega_0 - 1$. The light gray line corresponds to the linear case when $\lambda = 0$ (see also Figure 3.5), while the blue and green lines correspond to nonlinear cases ($\lambda > 0$ and $\lambda < 0$). For the nonlinear cases, we can see multi-modality and hysteresis: there is a segment of the resonance curve $a_0(\sigma)$ where two stable solutions of a_0 and one unstable a_0 (shown by the dashed line) coexist. For a color version of the figure, see www.iste.co.uk/basset/keh.zip*

Let us plot the nonlinear resonance curve. It represents a graph of a_0 as a function of the frequency ω_{ext} of the external vibrations. Alternatively, we can plot a_0 as a function of the mismatch σ between the natural frequency ω_0 and the external frequency ω_{ext} (the case $\sigma = 0$ would correspond to exact resonance). The nonlinear resonance curve of the Duffing oscillator is shown in Figure 6.1 (we have already seen this figure in Chapter 4). There are three cases shown in the figure: $\lambda = 0$ (corresponds to the resonance in a linear damped resonator), $\lambda < 0$ (nonlinear, with the spring softening effect) and $\lambda > 0$ (nonlinear, with the spring hardening effect). All of them are calculated for the same amplitude of the external force $\alpha = 0.2$. In contrast to the linear case shown by the gray line, resonance in nonlinear resonators leads to bi-modality (multi-modality) and hysteresis.

From this figure, we can see that for certain values of the frequency mismatch σ there are three overlapping branches of the graph. This means that there are three solutions for a_0 found from equation [6.9]. The further analysis of these solutions shows that two of them are stable (shown by the solid line), while the third is not stable (this branch is shown by the dashed line). Depending on initial conditions $x(t_0)$ and $v(t_0)$, the resonator can display forced vibrations with a large amplitude $a_{0,1}$ or with a smaller amplitude $a_{0,2}$. This phenomenon is called multi-modality.

Now assume that we can slowly change the frequency of the external oscillations ω_{ext} (i.e. we make a sweep of frequencies over the range from $-\omega_{ext}^{(start)}$ to $+\omega_{ext}^{(stop)}$). If we sweep the frequency of the external force in the direction shown by the arrows on the resonance curves, we will observe hysteresis. For example, let us consider the case $\lambda > 0$. Increasing the frequency ω_{ext}, we will observe gradual increase in the oscillation amplitude (arrow moving right in the figure). When the peak of the resonance curve is reached at some $\omega_{ext,1}$, we will observe a sharp jump as we will switch to the lower stable branch of the curve. Decreasing the frequency ω_{ext}, we will also observe gradual increase in the forced oscillations. At another frequency $\omega_{ext,2} < \omega_{ext,1}$, we observe a jump to a higher stable branch of oscillations. This phenomenon is known as the hysteresis of the nonlinear resonance curve.

Figure 6.2 shows the transformation of the resonance curve $a_0(\sigma)$ if the normalize amplitude of the external force α increases. If α is small enough, the system behaves similarly to a linear resonator and does not display any sign of bimodality or hysteresis. However, as the amplitude α of the external vibrations increases, the amplitude of forced oscillations also increases. The energy stored in the nonlinear resonator increases, and this changes the frequency associated with free oscillations of the nonlinear resonator (see our discussion on anisochronous oscillations and/or formula [5.34]). Thus, the peak of nonlinear resonance shifts to the right (to higher frequencies in the case of the spring hardening nonlinearity) or to the left (to smaller frequencies in the case of spring softening nonlinearities). In addition, the bandwidth of resonance response also increases. This is seen from a wider segment of bistabilitiy and hysteresis for larger α.

Therefore, nonlinear resonators can respond to external vibrations (maintain resonance, i.e. large amplitude oscillations) over a wide frequency range of ω_{ext}, which can be significantly larger than the "resonance" frequency range of a linear resonator. As we discussed in Chapter 4, this property is widely exploited in kinetic energy harvesters by designing and fabricating MEMS resonators with nonlinear springs. In this section, we showed the readers a rigorous way to describe this phenomenon.

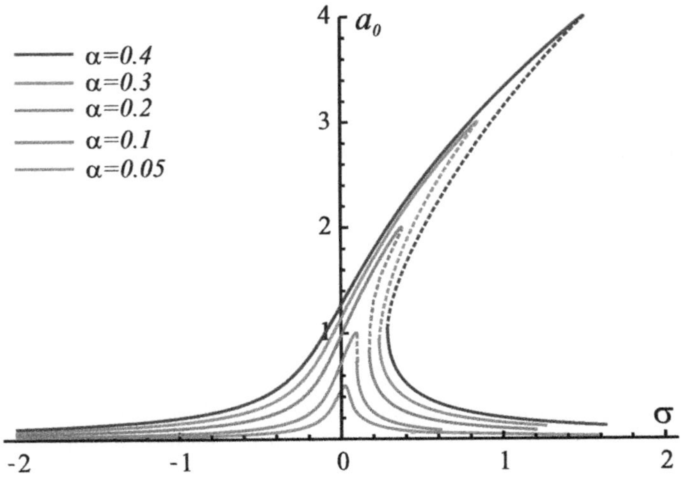

Figure 6.2. *Resonance in a nonlinear Duffing resonators with $\lambda > 0$. The larger the normalized amplitude of the external force α, the larger the amplitude of forced oscillations is. The segment of the resonance curve $a_0(\sigma)$ with bimodality and hysteresis (shown by the dashed lines) increases correspondingly. For a color version of the figure, see www.iste.co.uk/basset/keh.zip*

6.1.2. *Forced oscillations with a general form of nonlinear force*

From our discussion, the readers can see that we use the Duffing-type (cubic) nonlinearity to present and discuss the main effects produced by nonlinear forces on resonators. This is a convenient example of a nonlinear system. However, in Chapter 4, we mentioned that similar nonlinear effects can be caused by impacts, nonlinearities of a general polynomial type and even by the presence of the transducer force (strongly nonlinear term). Section 5.2.5 introduces an approach that can handle a variety of nonlinear functions. In this section, let us incorporate this approach to the analysis of forced nonlinear oscillations. We consider the equation:

$$y'' + 2\varepsilon\tilde{\beta}y' + y + \varepsilon\tilde{\lambda}G(y) = \tilde{\alpha}\cos((1+\varepsilon\tilde{\sigma})\tau + \phi_0) \qquad [6.11]$$

where we represent the nonlinear function using the Fourier series. In the case of high-Q resonators, we can limit the series by the first harmonic. And so:

$$\begin{aligned} G(y_0) &= G_0 + a_1(y_0)\cos(\tau+\varphi) + b_1(y_0)\sin(\tau+\varphi) \\ &= G_0 + c_1(y_0)e^{jT_0+j\varphi} + c.c. \end{aligned} \qquad [6.12]$$

Instead of equation [6.4], we obtain:

$$\begin{aligned} D_0^2 y_0 + y_0 &= 0 \\ D_0^2 y_1 + y_1 &= -2D_0 D_1 y_0 - 2\tilde{\beta} D_0 y_0 - \tilde{\lambda}\left[G_0 + c_1(y_0)e^{jT_0 + j\varphi} + c.c.\right] \\ &+ \left[(\tilde{\alpha}/2)e^{j((1+\varepsilon\tilde{\sigma})T_0 + \phi_0)} + c.c.\right] \end{aligned} \quad [6.13]$$

And so by setting the secular terms to zero, we obtain an equation on the complex amplitude A:

$$-2j\dot{A}e^{jT_0} - 2j\tilde{\beta}Ae^{jT_0} - \tilde{\lambda}c_1(y_0)e^{jT_0 + j\varphi} + (\tilde{\alpha}/2)e^{j\varepsilon\tilde{\sigma}T_0 + j\phi_0}e^{jT_0} = 0 \quad [6.14]$$

where the nonlinear force is represented by the complex coefficient c_1 that depends on the amplitude of oscillations. We will omit the intermediate steps of the method and provide the equations for the evolution of the amplitude and phase:

$$\begin{aligned} \dot{a} &= -\tilde{\beta}a + \frac{\tilde{\lambda}b_1}{2} + \frac{\tilde{\alpha}}{2}\sin\psi \\ \dot{\psi} &= \tilde{\sigma} - \frac{\tilde{\lambda}a_1}{2a} + \frac{\tilde{\alpha}}{2a}\cos\psi \end{aligned} \quad [6.15]$$

Therefore, the steady-state solution of the amplitude is

$$\left(\beta a_0 - \frac{\lambda b_1(a_0)}{2}\right)^2 + \left(\sigma a_0 - \frac{\lambda a_1(a_0)}{2}\right)^2 = \frac{\alpha^2}{4} \quad [6.16]$$

In order to apply this formula, we just need to compute the Fourier coefficient a_1 and b_1 of the general nonlinear force.

Finally, we will discuss the physical meaning of the components a_1 and b_1 of the force in this equation. Let us refer to the discussion in Chapter 3 on the nature of the forces. In the context of applications with oscillatory processes (periodically varying displacements and velocities), we can represent forces using their first harmonic, i.e. as the superposition of the sine and cosine components. We mentioned that it is sometimes interpreted as the superposition of potential and purely dissipative components. In order to investigate this, let us look at the effects that are produced by the components $a_1\cos(\tau + \varphi)$ and $b_1\sin(\tau + \varphi)$ of the force.

Assume that we want to find the frequency of the peak on the nonlinear resonance curve, i.e. we want to find at what frequency ω_{ext}^* of the external vibrations we observe the maximum amplitude a_0. We look for the extremum of the resonance curve equation, and so we compute the derivative of expression [6.16] with respect to ω_{ext}. Recall that we analyze the dimensionless case. We will discuss nonlinear resonance in terms of the frequency mismatch $\sigma = \omega_{ext}/\omega_0 - 1$, and so the derivative of the function in [6.16] with respect to σ is:

$$\frac{da}{d\sigma} = \frac{(a_0\sigma + \frac{a_1(a_0)}{2})a_0}{2a_0\beta^2 + \beta b_1(a_0) + \beta a_0 \frac{\partial b_1(a_0)}{\partial a_0} + \frac{b_1(a_0)}{2}\frac{\partial b_1(a_0)}{\partial a_0}} \quad [6.17]$$

Setting this equal to 0 to find its extremum, the resulting formula for the peak frequency shift σ^* is:

$$\sigma^* = -\frac{a_1(a_{max})}{2a_{max}} \quad [6.18]$$

and therefore the peak of the resonance curve is reached at the external frequency:

$$\omega_{ext}^* = \omega_0(1+\sigma^*) = \omega_0\left(1 - \frac{a_1(a_{max})}{2a_{max}}\right) \quad [6.19]$$

Note here again that the peak of the nonlinear resonance curve is not located at some fixed value of the external frequency ω_{ext} (in contrast, in linear systems, the resonance peak is always reached at $\omega_{ext} = \omega_0\sqrt{1-2\gamma^2/\omega_0^2}$). Formula [6.19] indicates that this peak explicitly depends on the amplitude of oscillations in the nonlinear resonator (i.e. on the energy stored in the resonator). As we have also seen from Figure 6.2, this peak is reached at different ω_{ext}^* for different values of the amplitudes a_0 of resonator oscillations. Thus, we have obtained another way to express the statement that nonlinear oscillations are anisochronous.

Moreover, from formula [6.19], we see that the component $a_1\cos(\tau+\varphi)$ of the nonlinear term $G(y)$ is responsible for this shift as it modifies ω_{ext}^*. And so we conclude that the cosine component of the force has the same effect as a purely potential force would have on this system. We also note that [6.19] is a useful formula as it allows us to quantify the effect of the frequency shift in nonlinear resonators.

With regard to the effect of the $b_1\sin(\tau+\varphi)$ component, it appears in the brackets with the normalized dissipation coefficient β and contributes to dissipation. Thus, depending on the sign and the particular form of b_1, it can decrease the energy of a resonator (introduces additional dissipation) or increase it. We classify such an effect as dissipative. As a final comment, note that the nonlinear cubic term $G(y) = y^3$ has only the potential component a_1 and does not have the dissipative component b_1. Since it is a purely potential force, it causes only the shift of the oscillation frequency and does not affect or introduce energy dissipation in the resonator (as expected).

6.2. Electromechanical analysis of an electrostatic kinetic energy harvester

6.2.1. *Statement of the problem*

Let us apply our knowledge of nonlinear resonators to the analysis of energy harvesters. We start by stating the equation that governs the evolution of the harvester's mechanical components (resonator). The dimensional displacement x of the KEH resonator is driven by external vibrations and affected by the transducer force, as discussed in Chapter 3. It can be described using the following resonator equation:

$$m\ddot{x} + \mu\dot{x} + kx = mA_{ext}\cos(\omega_{ext}t + \theta_0) + F_t(x, \dot{x}) \quad [6.20]$$

where m is the mass of the resonator, μ is the damping factor, k is the linear spring constant, ω_{ext} is the external frequency, θ_0 is the initial phase of the external vibrations, A_{ext} is the external acceleration amplitude and $\omega_0 = \sqrt{k/m}$ is the natural frequency of the linear resonator. Note that the inclusion of the force F_t makes the equation nonlinear in the most general case. We can extend this equation and include mechanical nonlinearities of the form $k_n x^n$ $(n > 1)$:

$$m\ddot{x} + \mu\dot{x} + kx + \sum_{n=2}^{N} k_n x^n = mA_{ext}\cos(\omega_{ext}t + \theta_0) + F_t(x, \dot{x}) \quad [6.21]$$

The function F_t is the transducer force and was defined previously. We give it here for the sake of completeness:

$$F_t = \frac{V_t^2}{2}\frac{dC_t}{dx} \quad [6.22]$$

The formula for the transducer force [6.22] requires to define:

1) The *capacitance* C_t as a function of the displacement x. The gradient of the capacitance dC_t/dx is used to determine the force. From this point, it becomes clear that the transducer force strongly depends on the geometry of the variable capacitor. Two variable capacitors with different C_t will have different dependances of the force on the displacement even if they operate with the same conditioning circuit. Please refer to Chapter 2 on the capacitive transducers for a detailed discussion on the capacitances-displacement characteristic.

There are a number of common capacitor configurations. For example, the symmetrical gap closing (SGC) capacitor is characterized by the following function:

$$C_t(x) = \frac{C_0}{[1 - x(t)/d]} + \frac{C_0}{[1 + x(t)/d]} = \frac{2C_0}{[1 - x^2(t)/d^2]} \qquad [6.23]$$

where C_0 is the capacitance of the transducer at rest, with no voltage applied, and d is the gap at rest.

The non-symmetrical gap closing (GC) capacitor is:

$$C_t(x) = \frac{C_0}{[1 - x(t)/d]} \qquad [6.24]$$

where C_0 and d are the same quantities as described before.

In the literature, we can also meet the area-overlap (AO) capacitor:

$$C_t(x) = C_0(1 + \alpha_c x(t)) \qquad [6.25]$$

where C_0 is again the capacitance at rest and α_c is an AO capacitor parameter. The capacitance of the AO capacitor changes linearly with the displacement.

2) The *configuration of the conditioning circuit*. Indeed, the voltage V_t appearing in equation [6.22] is not constant, and it is defined by the architecture of a conditioning circuit. There are a number of circuits used with capacitive transducers. They can be divided into three families: constant charge (or constant voltage) [MUR 04, MEN 01, MIT 12], continuous [MIT 08, HAL 08, BAS 14] and rectangular cycle [YEN 06, ROU 02, GAL 15] circuits. We would refer our readers to Chapter 8 on conditioning circuits for a more detailed discussion. Let us highlight that even if the same capacitive transducer operates with different conditioning circuits, it gives rise to different nonlinear forces F_t and, thus, to different dynamics of the resonator.

In order to conduct the electromechanical analysis employing equations [6.20] or [6.21], we must obtain the expressions for F_t in the explicit form, either from inspecting the QV-cycle or knowing the topology of the circuit. As our main example, we will discuss the electromechanical analysis of the constant charge circuit. However, we will provide examples of other conditioning circuit families. The method was adopted and used for e-KEHs in study [BLO 13], including some refinements that allow a more accurate estimation of the zeroth harmonic (the DC offset).

6.2.2. *Mathematical model of the constant charge circuit*

Let us start with a circuit implementing one of the fundamental energy conversion cycles (constant charge cycle), whose schematic diagram is shown in Figure 6.3. We will discuss the constant charge circuit in detail since it is a simple and suitable example to demonstrate the MSM method and discuss results.

The conditioning circuit discharges the transducer to zero when the transducer capacitance is at a local minimum and charges it to a charge Q_0 when its capacitance is at a local maximum. The energy conversion is achieved when the transducer capacitance decreases, keeping its charge constant (Q_0). During this process, mechanical energy is converted into electrical energy, and the transducer acts as a damper in the mechanical domain. In the case of transducers with monotonously increasing $C_t(x)$ characteristics, the voltage generated by the transducer depends on the sign of the mobile mass velocity: $V_t = 0$ if $\dot{x} > 0$ and $V_t = Q_0 C_t(x(t))$ if $\dot{x} < 0$. Hence, the force is a piecewise defined function: $F_t = 0$ if $\dot{x} > 0$ and $F_t = F_t(x, Q_0/C_t(x))$ otherwise, and will be referred later as $F_t(x, \dot{x})$.

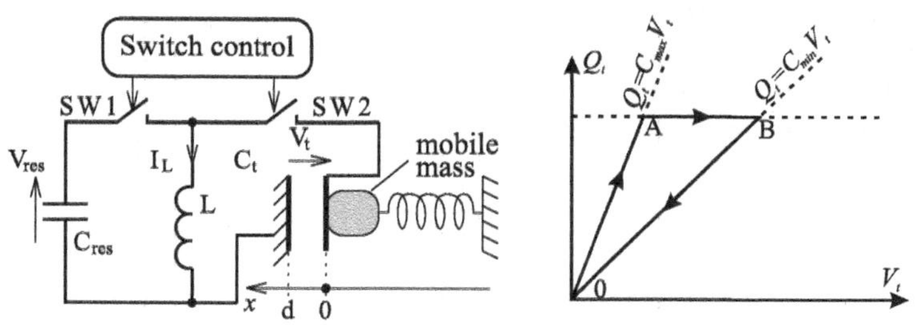

(a) (b)

Figure 6.3. *Schematic diagrams of the constant charge a) and charge pump b) circuit families*

At a local maximum of C_t, the conditioning circuit fixes three electrical quantities on the transducer: the charge Q_0, the voltage V_0 and the energy W_0. Only one of the three can be fixed independently of the others since they are related by the following expressions:

$$Q_0 = C_{max} V_0 , \qquad W_0 = \frac{1}{2} \frac{Q_0^2}{C_{max}} \tag{6.26}$$

Whatever quantity is fixed on the transducer at a local maximum of C_t, the charge Q_0 does not change until the next local minimum of C_t is reached. The electromechanical energy conversion is carried out during the time interval corresponding to the motion of the mobile plate from C_{max} to C_{min} positions. This energy conversion mode is called in literature the constant-charge operating mode of the transducer, which underlines the fact the transducer keeps a constant electrical charge during the electrical energy generation.

At a local minimum, the three quantities are set to zero. The quantity (one from the three) that is independently fixed to a non-zero value at a local maximum depends on the architecture of the conditioning circuit. In this chapter, for convenience, we consider the case when the energy W_0 is fixed. The behavior of the system with V_0 fixed is very similar and the equivalent voltage V_0 can be computed using equation [6.26].

Thus, using that $V_t = \sqrt{2W_0 C_{max}}/C_t$, we find:

$$F_t = \begin{cases} \frac{W_0 C_{max}}{C_t^2} \cdot \frac{dC_t}{dx}, & v \leq 0 \\ 0 & v > 0 \end{cases} \qquad [6.27]$$

Note that you can write this formula with respect to any quantity: W_0, V_0 or Q_0. Substituting a particular capacitance gradient dC_t/dx (for SGC, GC or AO transducer) will complete the definition of the force. For example, taking the GC transducer [6.24], we obtain the following expression:

$$F_t = \begin{cases} \frac{W_0}{d(1-x_{max}/d)}, & v \leq 0 \\ 0 & v > 0 \end{cases} \qquad [6.28]$$

while taking the AO transducer [6.25], we obtain

$$F_t = \begin{cases} \frac{W_0 \alpha_c (1+\alpha_c x_{max}/C_0)}{C_0(1+\alpha_c x/C_0)^2}, & v \leq 0 \\ 0 & v > 0 \end{cases} \qquad [6.29]$$

Here, x_{max} is the local maximum of x, defined similarly to C_{max}. We have also taken into account that $V_t = Q_0/C_t = \sqrt{2W_0 C_{max}}/C_t$. The expression for the transducer force is used directly in the mass-spring-damper equation [6.20]. We will see that despite the fact that F_t depends on x_{max} and is piecewise-defined, it is easy to incorporate this force in the MSM analysis.

Finally, we note that in order to apply the MSM, we should normalize equations [6.20]–[6.29]. For example, for the GC transducer, we select the following normalized variables and parameters: normalized displacement $y = x/d$, time $\tau = \omega_0 t$, dissipation $\beta = \mu/(2m\omega_0)$, external frequency $\Omega = \omega_{ext}/\omega_0 = 1 + \sigma$, external acceleration amplitude $\alpha = A_{ext}/(d\omega_0^2)$, normalized force coefficient

$\nu_W = W_0/(d^2 m\omega_0^2)$ and mechanical nonlinearity coefficients $\kappa_n = \frac{k_n d^{n-1}}{m\omega_0^2}$. This allows us to obtain the following normalized equations:

$$y'' + 2\beta y' + y + \sum_{n=2}^{N} \kappa_n y^n = \alpha \cos(\Omega\tau) + f_t(y, y') \qquad [6.30]$$

where f_t is the normalized transducer force

$$f_t(y, y') = \begin{cases} \frac{\nu_W}{1-y_{max}} & y' < 0 \\ 0 & y' > 0 \end{cases} \qquad [6.31]$$

6.2.3. *Steady-state nonlinear oscillations*

We apply the MSM method to equation [6.30] as discussed in section 5.2.4 taking into account the general form of nonlinear term $G = -f_t$ (note that sign of f_t as it stands on the right-hand side of the equation). First, we assume that the resonator is high-Q and therefore responds effectively only to the fundamental harmonic. This allows us to exclude higher harmonics from the response. The steady-state solution to the MSM equations [6.13] simplifies to:

$$y(\tau) = y_{av} + a_0 \cos(\Omega\tau + \phi_0 - \psi_0) \qquad [6.32]$$

where y_{av} is a possible constant shift that may appear if the nonlinear force f_t has a DC component:

$$y_{av} = f_0(a_0) \qquad [6.33]$$

Here, f_0 is the zeroth (DC) harmonic of f_t (see formulae [5.56]). The steady-state amplitude a_0 is found from equation [6.16]:

$$\left(\beta a_0 + \frac{b_1^{(t)}(a_0)}{2}\right)^2 + \left(\sigma a_0 + \frac{a_1^{(t)}(a_0)}{2}\right)^2 = \frac{\alpha^2}{4} \qquad [6.34]$$

If it is required, we can accommodate nonlinear terms $\sum k_n y^n$ employing the same approach. Therefore, in the most general case of the transducer force and mechanical nonlinearities, we obtain the following equation on the steady-state amplitude a_0:

$$\left(\beta a_0 + \frac{b_1^{(t)}(a_0)}{2} - \sum_{n=2}^{N} \frac{b_1^{(nl,n)}(a_0)}{2}\right)^2 + \left(\sigma a_0 + \frac{a_1^{(t)}(a_0)}{2} - \sum_{n=2}^{N} \frac{a_1^{(nl,n)}(a_0)}{2}\right)^2 = \frac{\alpha^2}{4} \qquad [6.35]$$

In [6.34] and [6.35], $a_1^{(t)}$ and $b_1^{(t)}$ are the Fourier coefficients of the fundamental harmonic of f_t, while $a_1^{(nl,n)}$ and $b_1^{(nl,n)}$ are the Fourier coefficients of the fundamental harmonics of the nonlinear terms $k_n y^n$. The coefficients $a_1^{(t)}$ and $b_1^{(t)}$ are found by substituting the expressions for the transducer force f_t to equation [5.56].

For instance, in the case of the gap-closing transducer with f_t defined as [6.28] (or its dimensionless analog [6.31]), the coefficients of the zeroth and first Fourier harmonics are:

$$\begin{aligned} &f_0(a_0) = \tfrac{\nu_W}{2(1-a_0)} \\ &a_1^{(t)}(a_0) = 0, \quad b_1^{(t)}(a_0) = \tfrac{2\nu_W}{\pi(1-a_0)} \end{aligned} \qquad [6.36]$$

For the area-overlap transducer with f_t defined by [6.29], the coefficients of the Fourier series are:

$$\begin{aligned} &f_0(a_0) = \tfrac{\kappa_0}{2(1-a_0)\sqrt{1-a_0^2}} \\ &a_1^{(t)}(a_0) = -\tfrac{\kappa_0 a_0}{(1-a_0)(1-a_0^2)^{1/2}}, \quad b_1^{(t)}(a_0) = \tfrac{2\kappa_0}{\pi(1-a_0)} \end{aligned} \qquad [6.37]$$

where $\kappa_0 = \alpha_c^2 W_0/(C_0^2 m\omega_0^2)$.

These coefficients are substituted into [6.34] to solve for a_0. As soon as the steady-state amplitude is found, we are able to write the displacement y in the form of oscillation [6.32]. If required for the further analysis, we can express $C_{max} = C_t(y)|_{y=y_{av}+a_0}$ and $C_{min} = C_t(y)|_{y=y_{av}-a_0}$ and compute the converted power:

$$P_{converted} = W_0 \left(\frac{C_{max}}{C_{min}} - 1\right) f_{ext} \qquad [6.38]$$

This allows the complete optimization of an e-KEH employing the constant-charge circuit [BLO 12].

For the simulation of a KEH with the constant-charge conditioning circuit, we have used the set of parameters given in Table 6.1. We are going to present the results of simulations of a mechanically linear resonator (i.e. nonlinearity is introduced only by the transducer force F_t), but mechanical nonlinearities can be taken into account as we discussed before. In order to investigate the nonlinear response of the harvester to external vibrations, we compute the steady-state amplitude a_0 from [6.34] and plot it as a function of the external frequency $f_{ext} = \omega_{ext}/(2\pi)$. The accurate comparison of the MSM results with direct solution of the model equations and with circuit simulations has been carried out [BLO 13, O'RI 14] and it has been shown that the MSM provides correct results. Therefore, while presenting the steady-state oscillations, we will give only the MSM results.

Resonator:	
Mass m	$6.35 \cdot 10^{-5}$ kg
Spring constant k	100 Nm^{-1}
Natural frequency ω_0 and f_0	1254.9 rad and 199.7 Hz
Quality factor Q	20
Location of stoppers	36.0 μm
AO transducer:	
Area S_{AO}	$3.50144 \cdot 10^{-4}$ m^2
Transducer coefficient α_c	$6.667 \cdot 10^{-7}$ Fm^{-1}
GC and SGC transducers:	
Area S_{GC}	$1 \cdot 10^{-4}$ m^2
Rest gap d	$45 \cdot 10^{-6}$ m
Other parameters:	
Acceleration amplitude A_{ext}	$0.25g$ m/s^2
External frequency ω_{ext}	170–220 Hz
Energy applied to the transducer W_0	0.5–10 nJ
Bias voltage V_0	5–40 V

Table 6.1. *Parameters of the resonator, transducer and conditioning circuit for the simulation of the constant-charge conditioning circuit*

The frequency sweeps $a_0(f_{ext})$ for the KEH of Table 6.1 are presented in Figure 6.4. In these figures, we demonstrate (1) the effect of biasing by increasing W_0 and thus increasing the electromechanical coupling and (2) the effect of different transducer by considering AO, GC and SGC transducers. It is interesting to highlight that in the case of the GC transducer, formula [6.36] yields $a_1 = 0$ and $b_1 \neq 0$. This indicates that the transducer force acts as a purely dissipative force, reducing the energy in the resonator and not contributing to the shift of the resonance frequency. This result is shown in Figure 6.4(a) where we increase the biasing of the resonator (through increasing W_0). The peak of the resonance curve denoted as f_e occurs around the same frequency for all biasing, i.e. $f_{e,1} = f_{e,2} = f_{e,3}$. At the same time, increasing biasing clearly reduces the resonator energy, as the peak value a_{max} decreases. The transducer force acts similarly to Coulomb's damping [AND 87].

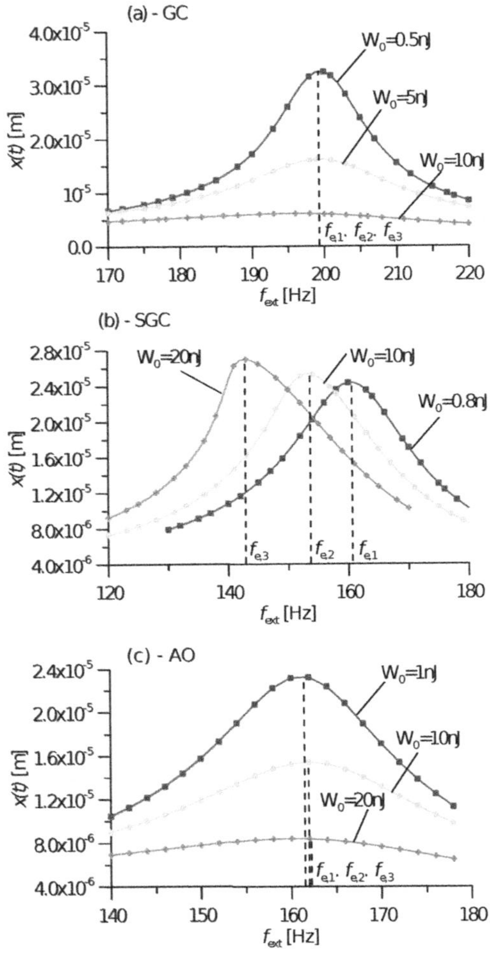

Figure 6.4. *Frequency sweep $a_0(f_{ext})$ showing the nonlinear resonance curve for a KEH employing a constant charge circuit. The parameters of the resonator, transducer, circuit and input vibrations are given in Table 6.1. Three figures correspond to the following geometries of the transducer: a) gap-closing, b) symmetrical gap-closing and c) area-overlap. Each figure illustrates the effect of biasing and electromechanical coupling by showing three graphs with increasing W_0. For a color version of the figure, see www.iste.co.uk/basset/keh.zip*

In the case of the AO transducer, formula [6.37] indicates us that both components $a_1 \neq 0$ and $b_1 \neq 0$. The coefficient $a_1 < 0$, and it causes the resonant peak to shift to larger frequencies $f > f_0 \simeq 200$ Hz according to [6.19]. This is shown in Figure 6.4(c) where we observe a "spring hardening effect", similar to that of nonlinear resonators with cubic or other hardening mechanical nonlinearities. It is termed *electrostatic hardening* to highlight that the nonlinearity that is responsible for this effect comes from the electrical domain. However, an increase in biasing and electromechanical coupling causes, as in the previous case, a dissipative effect.

For the SGC transducer, the coefficient $a_1 > 0$ (not shown in this chapter), and it causes the resonant peak to shift to smaller frequencies $f < f_0 \simeq 200$ Hz. This situation is shown in Figure 6.4(b). This "spring softening effect", similar to that of nonlinear resonators with quadratic or other softening mechanical nonlinearities, is termed *electrostatic softening*. Another interesting effect is that an increase in biasing does not reduce the energy in the resonator as in the case with the GC transducer. On the contrary, the peak value a_{max} increases.

Whether a_{max} increases or decreases with biasing depends on a particular combination of a_1 and b_1. We want to stress again that a particular combination of a conditioning circuit and a transducer may lead to all possible effects: electrostatic hardening or electrostatic softening, reducing the energy in the resonator (purely dissipative action) or increasing it (negative dissipation effect). We will see from later examples of other conditioning circuits that, indeed, all these effects are possible.

6.2.4. *Dynamical effects and bifurcation behavior*

While the MSM analysis allows us to find steady-state quasi-harmonic oscillations in an e-KEH, it also allows us to study for its stability. Although we will not discuss stability analysis in details here, we will show in this section that, similarly to other nonlinear resonators such as the Duffing resonator, KEHs can display bifurcations and chaotic modes of oscillations [HIL 00, NAY 08a, HIR 12].

A simple way to show the nonlinear effects in an KEH is by applying a slowly growing ramp of acceleration amplitude A_{ext} to the system and investigating the system's response to this changing acceleration amplitude using the original, normalized, mathematical model [6.20]. This process can be considered to be quasi-static since the ramp of A_{ext} changes very slowly. It is clear from Figure 6.5 that the system's dynamics undergo various changes as a result of varying A_{ext}. Figure 6.5 can be seen as an analog of a bifurcation diagram with A_{ext} being the bifurcation parameter. By examining Figure 6.5, the following qualitative changes (bifurcations) can be seen, as a result of changing A_{ext}: (1) the appearance of steady-state harmonic oscillations (fragment 1 in the figure) and (2) period-doubling bifurcation (fragment 2) and (3) transition to chaos (fragment 3).

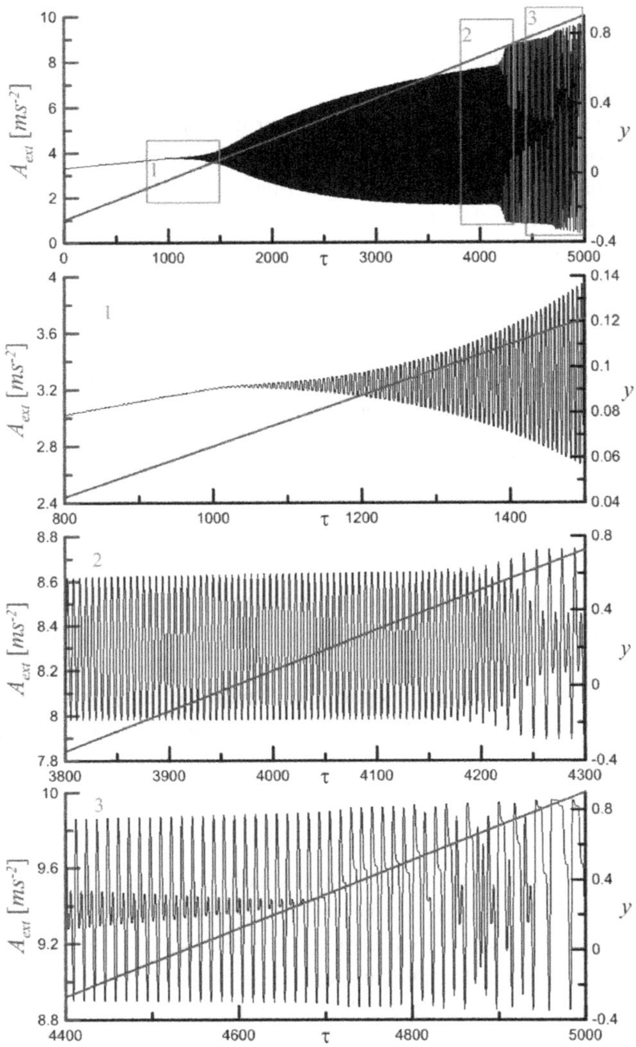

Figure 6.5. *The response of an e-KEH with a constant charge circuit to a growing external acceleration amplitude. In the top subfigure, a slowly growing ramp of A_{ext} can be seen along with the normalized resonator displacement y. The different changes that the system's dynamics undergo have been highlighted and magnified. In fragment 1, the appearance of steady-state harmonic oscillations can be seen. In fragment 2, period doubling bifurcation can be seen; and in fragment 3, the transition to chaos can be seen. This figure is analogous to a bifurcation diagram*

In order to obtain a more global picture of the dynamics of an e-KEH using numerical simulations of [6.30], it is possible to construct what is called a *bifurcation diagram.* A bifurcation diagram is constructed by fixing the values of all but one of the control parameters of the system. The control parameter that is varied is called the bifurcation parameter. Initially, [6.30] is simulated for a specific value of this bifurcation parameter along with a particular set of initial conditions. Disregarding the transient, a Poincaré section of the system is taken periodically with period $T = 2\pi/\Omega$. This gives values for the normalized displacement and velocity of the system (y_1, y_1') at every period. If we are operating in a steady-state period-1 regime, then the trajectory will cross the Poincaré section at (y_1, y_1') after every T units of dimensionless time. If we are operating in a steady-state period-2 regime, then the trajectory will cross the Poincaré section at (y_1, y_1') after every $2T$ units of dimensionless time. If it is a steady-state period-4 behavior, then the values of displacement and velocity will be (y_1, y_1') after every $4T$ units of dimensionless time and so on. Using the values obtained through the numerical simulations and Poincaré mapping, we can plot the values for y or y' against the bifurcation parameter. The value of the bifurcation parameter is then changed and the simulations redone using the same initial conditions. This process is repeated for a chosen range of the bifurcation parameter. The resulting plot, the bifurcation diagram for the system, although very time-consuming to produce, gives a lot of very useful information about the global dynamics of the system, regions of steady-state behavior, bifurcation points and tells us how the system transitions to chaos. An example of such a bifurcation diagram for an e-KEH can be seen in Figure 6.6. In this instance, the acceleration of the external vibrations, A_{ext}, is the chosen bifurcation parameter and it is the evolution of the normalized displacement y that is being observed as A_{ext} is being varied. It is clear to see the benefits of numerically constructed bifurcation diagrams from Figure 6.6. A quick glance at the figure allows us to determine the global dynamical behavior of the system, how it transitions to chaos and approximate bifurcation points.

6.2.5. *Other conditioning circuits*

We will briefly discuss the results of steady-state analyzing for two e-KEHs with circuits from other families: the continuous conditioning circuit and the charge pump circuit with resistive flyback. In contrast to the constant charge circuit, where we are able to write the dependance of the transducer force directly on the displacement x and velocity v, the analysis of these two cases requires some additional steps. The details on how these systems can be handled can be found in [GAL 13a, O'RI 15]. More details on conditioning circuits will be provided in later chapters. Here, however, we would like to state the mathematical models of these systems:

– *Continuous circuit*: the continuous conditioning circuit consists of a resonator with an electrostatic transducer connected in series to a resistive load R_L and some form of electric bias so as to place a charge on the capacitor. This may be in the form of a voltage source V_0, large charged capacitor or electret layer of the capacitive transducer: in the three cases, the circuit behavior is similar. The diagram of an e-KEH employing the continuous conditioning circuit is shown in Figure 6.7(a).

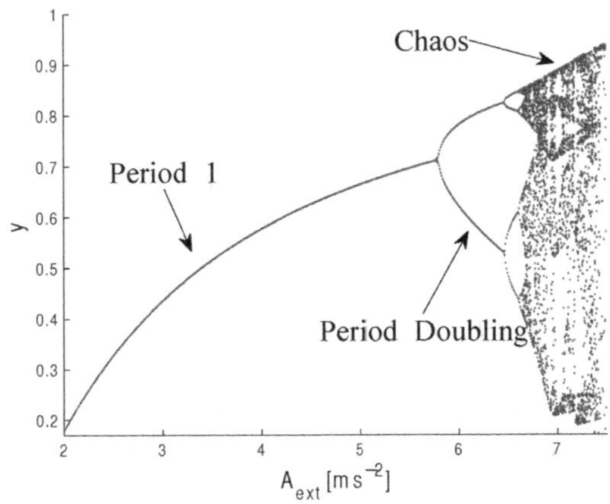

Figure 6.6. *Bifurcation diagram for an e-KEH with a constant charge circuit for $W_0 = 10nJ$. In this instance, the bifurcation parameter is A_{ext} and y is the parameter being plotted. The values of y are found through numerical integration with the same set of initial conditions $y = y' = 0$ for every value of A_{ext}. A bifurcation diagram like this gives a lot of information about the global dynamics of a system. Different regions of periodic behavior can be seen along with the system's transitioning to chaos. Windows of periodicity can even be seen within the chaos at higher values of A_{ext}*

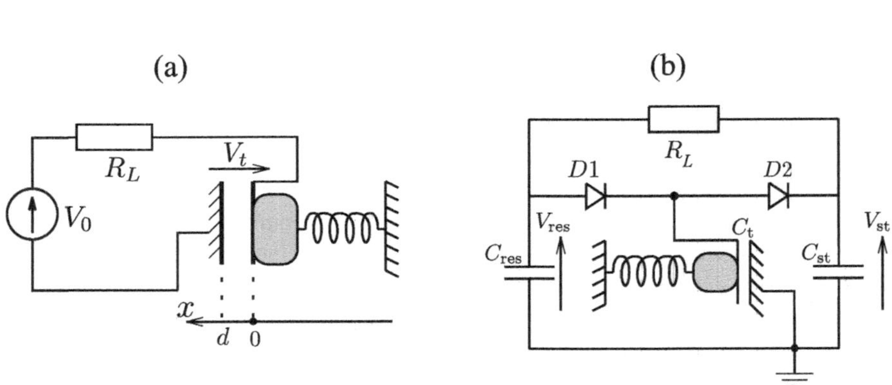

Figure 6.7. *Schematic diagram of the continuous a) and charge pump b) circuits*

The displacement x of the mobile mass with respect to the frame is described by the same equation as we encounter in the previous section, equation [6.20] with the transducer force defined by equation [6.22]. We want to highlight that these are universal equations valid for all e-KEHs. However, the governing equations describing the electrical behavior of the simple conditioning circuit are now different. By the Kirchhoff voltage law and the element equations, we obtain:

$$R_L \frac{dQ_t}{dt} + \frac{Q_t}{C_t(t)} = V_0, \qquad [6.39]$$

where Q_t is the instantaneous charge on the transducer capacitor and $C_t(t)$ is the time evolution law of the transducer's capacitance. In order to complete the set, we must give an equation to relate the charge Q_t with the transducer voltage V_t. This equation is:

$$V_t(t) = Q_t(t)/C_t(t). \qquad [6.40]$$

– *Charge pump circuit with resistive flyback*: an e-KEH employing the charge pump circuit is shown in Figure 6.7(b). The voltage source V_{res} is usually implemented with a large reservoir capacitor, whose value is large enough to neglect its voltage variation over one capacitance variation period. The voltage source V_{st} is implemented by a storage capacitor C_{st}, which is small compared to C_{res}, but much greater than the maximum value of the transducer capacitance C_{max}. C_t capacitor is initially charged to a voltage $V_{t,0}$, and so are C_{st} and C_t capacitors, because of the charge sharing through the diodes D1 and D2. Making the use of diode models, capacitor models and applying Kirchoff's circuit laws, the system in the electrical domain can be reduced to three differential equations describing the voltages V_{res}, V_t, V_{st}:

$$\begin{aligned} \dot{V}_{st} &= \tfrac{1}{C_{st}}\left(K(-V_t - V_{st}) - \tfrac{(V_{st}-V_{res})}{R}\right) \\ \dot{V}_{res} &= \tfrac{1}{C_{res}}\left(\tfrac{(V_{st}-V_{res})}{R} - K(V_{res} + V_t)\right) \\ \dot{V}_t &= \tfrac{1}{C_t}\left((K(-V_t - V_{st}) - K(V_{res} + V_t)) - V_t\, \dot{x}\, \tfrac{dC_t}{dx}\right) \end{aligned} \qquad [6.41]$$

where K is the diode model current $K(V) = I_s(e^{qV/kT} - 1)$, I_s is the saturation current, q is the electron charge, k is Boltzmann constant and T is absolute temperature.

For both systems, the amplitude of steady-state oscillations is calculated from [6.34]. The parameters we used in simulations are given in Table 6.1. The frequency sweeps $a_0(f_{ext})$ for a KEH with the continuous conditioning circuit and with the charge pump circuit are shown in Figure 6.8 (the left and right columns, respectively). Three graphs in each column correspond to the GC, SGC and AO transducer. Each graph also shows the effect of biasing (three different values of V_0). We note that the GC and SGC transducers produce the electrostatic softening effect if coupled with both, the continuous and charge pump circuits. In contrast, the AO transducer produces the electrostatic hardening effect. We also note that the force that is generated by the GC transducer has a dissipative effect in all the three circuits, including the continuous and charge pump ones. On the contrary, the force due to the SGC and AO transducers when biased with the continuous and charge pump circuits does not increase the dissipation of energy in the resonator.

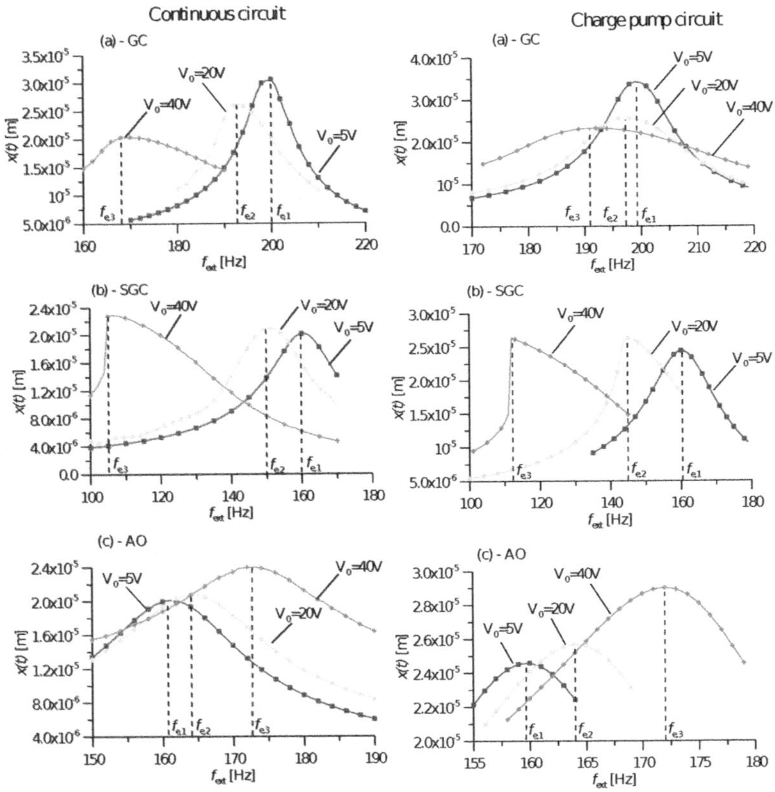

Figure 6.8. *Frequency sweep $a_0(f_{ext})$ showing the nonlinear resonance curve for a KEH employing a constant charge (left column) and a charge pump with resistive flyback (right column) circuit. The parameters of the resonator, transducer, circuit and input vibrations are given in Table 6.1. The subfigures correspond to the following geometries of the transducer: a) gap-closing, b) area-overlap and c) symmetrical gap-closing. Each figure illustrates the effect of basing and electromechanical coupling by showing three graphs with increasing V_0. For a color version of the figure, see www.iste.co.uk/basset/keh.zip*

Finally, we note that at large bias voltages $V_0 \geq 40V$, we observe hysteresis in the case of the SGC transducer (both circuits, Figure 6.8(b)). The presence of hysteresis is validated experimentally in [BAS 14, O'RI 15]. Since we present the study without additional mechanical nonlinearities, we conclude that these nonlinear effects appear in e-KEHs only due to the transducer force, i.e. due to the electromechanical coupling. The presence of additional mechanical nonlinearities will enhance these phenomena [HAR 14]. The e-KEH with the charge pump circuit also undergoes the same cascade of bifurcations and route to chaos as the KEH employing the constant charge circuit (see Figures 6.5 and 6.6).

6.3. Concluding remarks

The electromechanical coupling due to the transducer force is the cause of the nonlinearities in the system. The influences of the nonlinear coupling include:

– the resonant frequency of oscillations can shift due to the electromechanical coupling, this also effects the amplitude of oscillations in the resonator;

– the change in resonator dynamics such as a cascade of bifurcations.

It is necessary to gain an enhanced understanding of the entire system in order to truly optimize the energy harvested by these devices. Even at low electrostatic biasing, the maximum energy is not harvested at the resonance frequency. In general, there is no way to determine what shift (softening or hardening) may occur for different transducers and parameter values without modeling the nonlinear coupling. Therefore, we simply cannot afford to neglect the influence of the transducer force and the nonlinearities it introduces. The model resulting from the multiple-scales method can be relatively easily adapted to include mechanical nonlinearities.

7

MEMS Device Engineering for e-KEH

A great advantage of electrostatic transducers is that the electrostatic force remains constant if all of the device's physical dimensions are reduced, as discussed in section 2.3.2. For this reason, we say that the electrostatic transduction is particularly favorable for the miniaturization. A good way for this is to use silicon-based microtechnologies, commonly called MEMS for *microelectro mechanical systems*. Indeed, silicon is a very good mechanical material, especially if its crystalline form is used: silicon's Young modulus varies from 130 to 170 GPa depending on the crystal orientation and the direction of bending, and its fracture limit is up to 7 GPa [PET 82]. In addition, since MEMS are usually manufactured through batch fabrication processes, the cost per device can be very low if mass produced.

Section 7.1 describes the basics of the MEMS technologies used for e-KEH fabrication. In section 7.2, typical device geometries are presented and some design considerations are discussed. Finally, section 7.3 introduces electret-biased e-VEHs.

7.1. Silicon-based MEMS fabrication technologies

Processes for MEMS fabrication can be roughly divided into two categories: the bulk and the thin-film technologies. The silicon bulk technology appeared in the 1950s for the fabrication of pressure sensors based on silicon's piezoresistance [SMI 54]. It consists of etching a crystalline silicon wafer over several tens or even hundreds of micrometers. MEMS thin-film technologies were developed in the 1980s and are directly inspired from *integrated circuit* (IC) technology. The bulky silicon wafer is used as a *handle wafer* on top of which thin films are deposited whose thicknesses are from a few tens of nanometers to a few micrometers. At the end of this process, one of these layers, called the *sacrificial layer*, is removed in order to allow the motion of structures designed in the layers above. If in bulk technologies the movable parts are made of crystalline silicon, then the mechanical layer in thin film technologies is typically made of polysilicon.

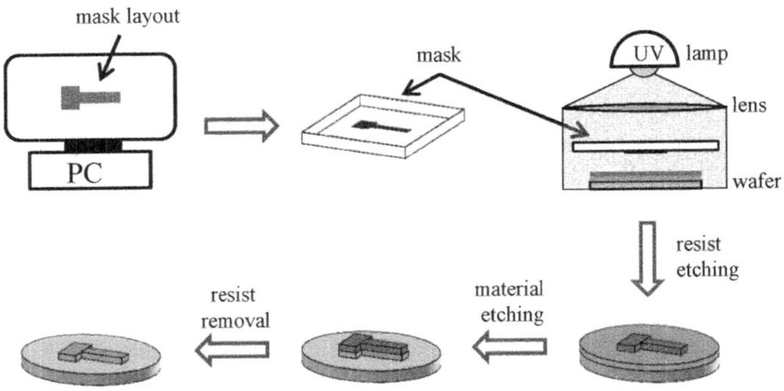

Figure 7.1. *Principle of photolithography in IC and MEMS technologies. For a color version of the figure, see www.iste.co.uk/basset/keh.zip*

MEMS technologies, similarly to ICs, are based on photolithography. This critical step is presented in Figure 7.1. For each material layer, the patterns to be etched are first designed using Computer Assisted Design (CAD) software. These patterns are then reproduced on a metal layer (usually chrome) deposited on a quartz plate due to e-beam lithography. This quartz plate is called a photomask and is partially transparent. It will serve as a template to insolate by ultraviolet (UV) light a photoresist layer deposited on top of the wafer to be etched. As such, UV light insolates the resist only where the chrome has been removed from the mask. This creates a chemical reaction in the resist such that it will be easy to remove, according to the mask patterns.

7.1.1. *Examples of bulk processes*

As can be seen from formula [3.37] which provides maximum power extractable from a resonator, an efficient e-KEH needs a heavy mass (m) which is able to move along a large distance (X_{lim}). For this perspective, bulk processes are probably the best techniques since silicon is a better mechanical material than the materials typically used in thin-film technologies. In addition, a relatively thick mass can be obtained without adding extra materials on top of the mobile part. The simplest way to obtain a bulk-silicon e-KEH is to use a *silicon on insulator* (SOI) wafer. SOI wafers are actually made up of two silicon wafers separated by a thin insulating layer, generally made of silicon dioxide (SiO_2). The bottom wafer, also called *handle wafer*, usually has no specific role except for supporting the top silicon wafer, the actual SOI layer. This kind of substrates was originally developed for high-frequency transistors, with a relatively thin SOI layer. However, although SOI wafers can be quite expensive compared to classical Si wafers, SOI layers with thicknesses up to 1 mm are commonly used for MEMS fabrication, with the buried insulator layer being used as the sacrificial layer.

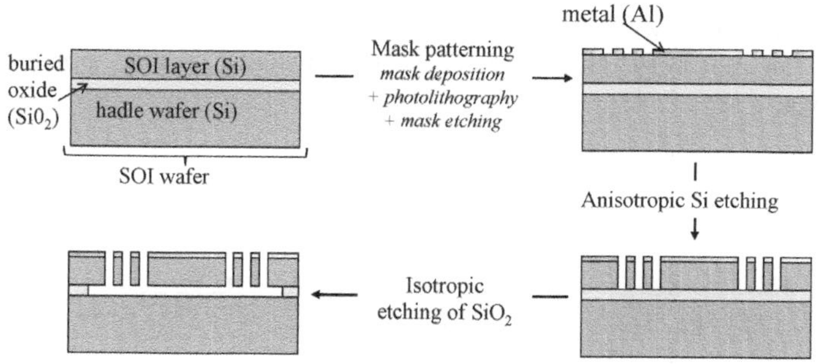

Figure 7.2. *Basic steps of MEMS fabrication using SOI wafers*

In the simplest case, only three steps and one photomask are required to fabricate a device, as illustrated in Figure 7.2. The first step consists of depositing a mask on top of the SOI layer, which is patterned by photolithography, and used as a protection layer for the silicon etching. Once the SOI layer has been vertically etched by *deep reactive ion etching* (DRIE), the buried oxide layer is isotropically etched by hydrofluoric acid.

With SOI wafers, the electrostatic transducer can only be made with vertical electrodes etched in the SOI layer. An alternative is to process two separate wafers and to bind them together. With this cheaper technology, the transducer's electrodes can be made in the same layer, or can be located on both substrates. Usually, a silicon wafer contains the mobile electrode(s) and the second wafer, which is not necessarily made of silicon, is part of the rigid body and supports the fixed electrode(s). Figure 7.3 represents a bulk process using a two-wafer technology (one silicon and one glass wafer). It requires two photomasks in order to obtain vertical electrodes in the silicon wafer [BAS 14].

7.1.2. *Thin-film technology with sacrificial layer*

In the thin-film process, thin layers of materials are deposited (or grown) on top of a silicon substrate. Movable parts are obtained by removing one of the layers at the end of the process. This way, the above layer(s) is (are) free to move, assuming that the mobile parts have been previously anchored to the substrate. The mechanical layer, i.e. the layer including the moving parts, is typically made of polysilicon and has a thickness ranging between 0.5 and 3 μm, although epitaxial techniques allow us to reach several tens of micrometers. Figure 7.4 illustrates the principle of the thin-film technology. Mechanical properties of polysilicon are not as good as those of crystalline silicon, which makes large displacements a possible cause of mechanical failure. In addition, the small thickness of the different layers significantly reduces the weight of the mobile mass compare to a KEH of the same area but made with a bulk process. It also reduces the transducer's capacitance ratio because of the high parasitic capacitance with the substrate. KEHs made of "pure" MEMS thin-film technology are not very common because they usually provide small amounts of output power.

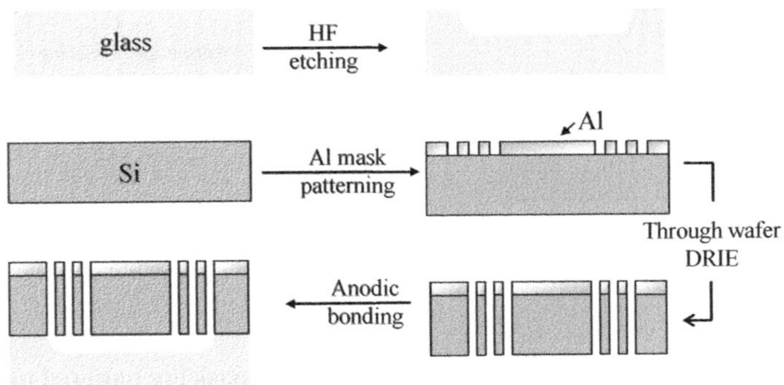

Figure 7.3. *Example of a silicon-glass process for MEMS [BAS 14]*

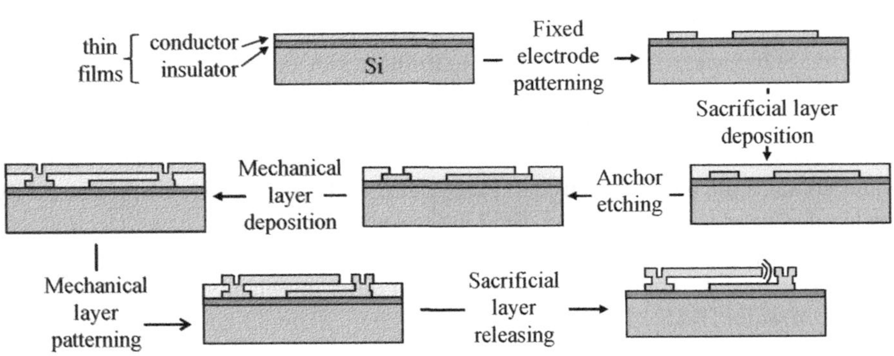

Figure 7.4. *Example of thin-film technology for MEMS*

7.2. Typical designs for the electrostatic transducer

The design of the e-KEH transducer has a strong impact on its efficiency and should be carefully chosen with respect to the expected external acceleration. e-KEH can be roughly divided into two categories: those whose transducer's capacitance variation originates from a change in the overlapping area of the electrodes, and those whose transducer's capacitance variation comes from a change in the gap between the electrodes (see discussion in section 2.1). With the first category, the capacitance usually varies linearly along the traveling of the mobile electrode. For the second category, a significant nonlinear increase in the capacitance is obtained if the acceleration is high enough to (almost) close the gap between the electrodes.

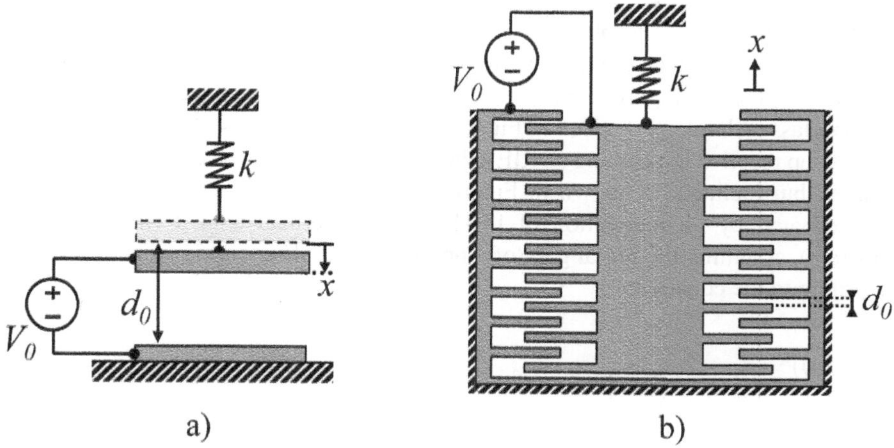

Figure 7.5. *Schematics of gap closing capacitive transducers: a) side view of a simple out-of-plane gap-closing (OPGC) geometry, b) top-view of an in-plane gap-closing (IPGC) geometry using interdigited comb-drives*

7.2.1. *Capacitive transducers with gap-closing electrode variation*

7.2.1.1. *Out-of-plane gap-closing (OPGC) transducers*

One of the easiest ways to obtain a variable capacitance in MEMS technology is to define two planar electrodes with an out-of-plane gap-closing scheme as illustrated in Figure 7.5(a). Assuming the fringe-field effects are negligible, the capacitance variation law is given by:

$$C_t = \varepsilon_0 \frac{S}{d_0 - x} \qquad [7.1]$$

where S and d_0 are the overlapped area and the initial gap of the electrodes, respectively, x is the displacement of the mobile electrode and ε_0 is the electric permittivity of the vacuum. The capacitance variation is inversely proportional to the mass displacement and can be very high, when the electrodes come in contact. To avoid electrical short-circuits, at least one of the electrodes has to be covered by an insulating layer and/or have mechanical stoppers located between the electrodes. Mechanical stoppers are also very practical to reduce the effect of pull-in and to limit the output voltage, as discussed in section 7.2.2.

7.2.1.2. *In-plane gap-closing (IPGC) transducers*

IPGC e-KEHs can be obtained with classical interdigited-comb geometry (see Figure 7.5(b)). Both electrodes are now in the same layer, which simplifies the fabrication process. Figure 7.6 shows an example of such a device using the bulk silicon-glass technology detailed in Figure 7.3. However, because of the undercut due to the deep reactive ion etching (DRIE), the plates of the capacitive transducer are not parallel, but inclined, as shown in Figure 7.6(b). The capacitance variation can be approximated by an integration of [7.1] over the vertical axis y (normal to the substrate), assuming the initial gap (of a non-deformed structure) is variable along y, according to the equation:

$$\delta(y) = d_0 + 2y\alpha, \qquad [7.2]$$

where d_0 is the gap between the fixed and movable fingers measured in the non-deformed device at the top of the comb and $\alpha = \delta/h_f$ is the ratio between the maximal lateral undercut δ and the height of the etched silicon h_f.

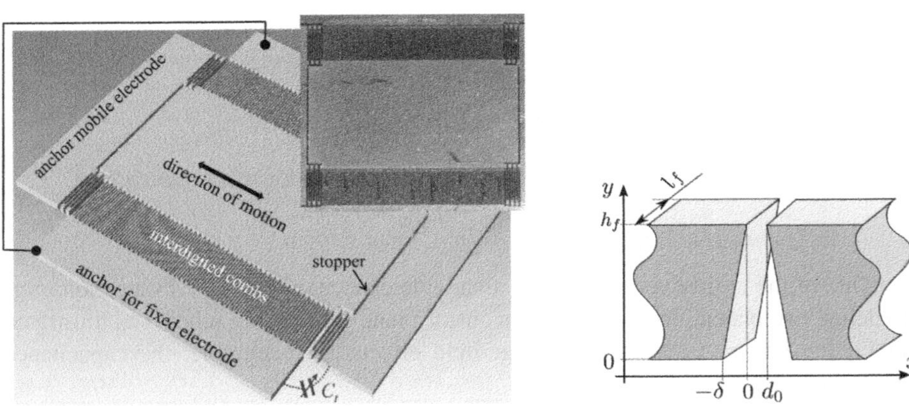

a) b)

Figure 7.6. *3D schematic a) and top-view photograph b) profile of the transducer's gap due to the DRIE etching of an in-plane gap-closing (IPGC) transducer [BAS 14]*

This yields the following equation [BAS 14]:

$$C_t(x) = N_f \left(\int_0^{h_f} \frac{\varepsilon_0 l_f dy}{d(y)-x} + \int_0^{h_f} \frac{\varepsilon_0 l_f dy}{d(y)+x} \right) \\ = \frac{\varepsilon_0 N_f l_f}{2\alpha} \cdot \ln \left[\frac{(d_0+2h_f\alpha)^2 - x^2}{{d_0}^2 - x^2} \right] \qquad [7.3]$$

where the sum in the parentheses is the capacitance between one movable finger and its two adjacent fixed fingers, N_f is the total number of movable fingers, l_f is the length of the overlapping part of the comb fingers and h_f is the thickness of the fingers (see Figure 7.6(b)). x is the instantaneous displacement of the mobile electrode (with regard to the non-deformed device shape). It assumes that the fringed field can be neglected, which is not the case if the comb's thickness is of the same order as the gap between the fixed and movable fingers. In this case, a correction factor needs to be added.

Equation [7.3] includes the parameter α to represent the slope during the comb etching. It is mandatory to take into account this parameter, which is directly related to the geometry of the pattern to be etched, the thickness of the comb fingers and the etching process. With DRIE of silicon, if a small initial gap d_0 gives a better aspect-ratio h_f/d_0, it also limits the depth of etching [PAR 14]. For instance, in [BAS 14], the etching of 400 μm of crystalline silicon with opening of about 40 μm led to α equal to 0.013. It had a dramatic influence on the capacitance variation, as shown in Figure 7.7.

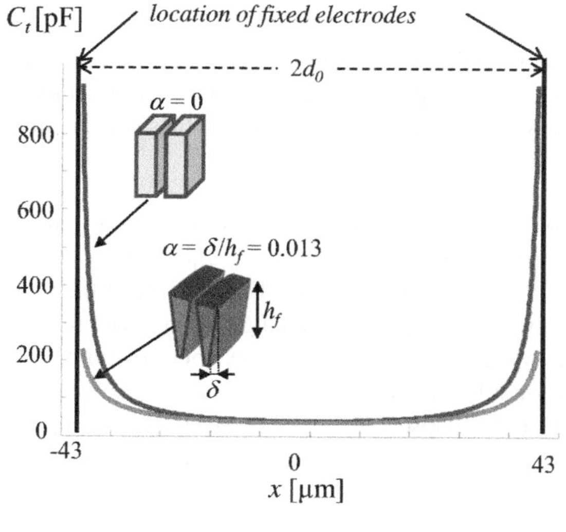

Figure 7.7. *Influence of the etching slope on the capacitance variation in gap-closing interdigited-comb transducers*

7.2.2. *Strategies on the stopper's location in gap-closing e-KEH*

Defining optimum values for the initial gap between the electrodes, the stopper size/location and the transducer's bias voltage is not trivial. For instance, if it is required to maximize the max-to-min ratio of the transducer's capacitance, the minimum capacitance should be minimized: for this, a large initial gap d_0 is needed, in association with low parasitic capacitances. However, a large gap also requests a high external acceleration to maximize C_t (to close the gap), which may not be available in the targeted application. In addition, a large d_0 increases the transducer's size, especially for the IPGC structure.

7.2.2.1. *Considerations on the trade-off between the bias and the displacement limit*

According to equation [3.37] and [2.30], the converted power from the mechanical domain is maximized for both a high bias voltage and a large mass displacement resulting in a large capacitance variation of the electrostatic transducer. A first limit for the bias is the static pull-in voltage V_{pi}. Above V_{pi}, there is no equilibrium between the restoring force of the spring and the electrostatic force, so that the mobile electrode collapses definitely toward the fixed electrode and no capacitance variation occurs anymore. For a parallel-plate capacitor described by [7.1], V_{pi} is given by:

$$V_{pi} = \sqrt{\frac{8kd_0^3}{27\varepsilon_0 S}}, \tag{7.4}$$

where k is the stiffness of the spring attached to the mobile electrode of the transducer. This voltage is called "static pull-in voltage" since it is measured or calculated on a structure without any external excitation, where the bias voltage increased slowly from 0 to V_{pi}.

For biases below V_{pi}, there are two equilibrium locations, one stable x_{stable} close to the mobile electrode and one unstable $x_{unstable}$ close to the fixed one, such that $0 < x_{stable} < d_0/3 < x_{unstable} < d_0$ (see Figure 7.8). If the mobile electrode is in motion (and hence, we are not in a static configuration anymore), the mobile electrode should absolutely avoid going beyond this unstable position: if it does, the spring restoring force may not be high enough to compensate the electrostatic force and the mobile electrode would collapse to the fixed one instead of backing off toward x_{stable}. Such a phenomena is called "dynamic pull-in", it may happen even for voltages less than V_{pi}.

The values of x_{stable} and $x_{unstable}$ are the roots of the following third-order polynomial equation corresponding to the equalization of the spring force and the electrostatic force of the transducer:

$$kx = \frac{1}{2}{V_0}^2 \frac{\varepsilon_0 S}{(d_0 - x)^2} \quad [7.5]$$

As far as $V_0 < V_{pi}$, this equation has three real roots. One of them corresponds to a position above d_0 and hence is not physical. Two others are positive and correspond to the equilibrium positions of the mobile electrode of the transducer. They are close to 0 and to d_0 for small V_0, and as V_0 increases they move closer to each other, ending up at $d_0/3$ when $V_0 = V_{pi}$, see Figure 7.8. As a result, $x_{unstable}$ reduces when V_0 increases; its value defines the maximum allowable furthest position of the mobile electrode along the axis x ($x_{\max}$), and hence, the value of $C_{\max}$.

Equation [7.5] is equivalent to a third-order polynomial equation, whose roots can be expressed analytically by using the Cardano's method [KOR 00]. However, the corresponding expressions are burdensome (they include cubic and square roots) and are seldom useful in practice.

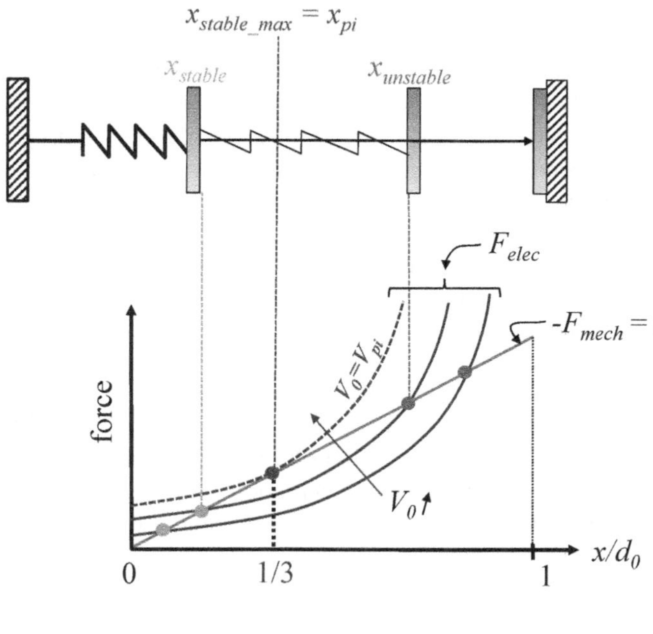

Figure 7.8. *Equilibrium positions for non-null bias voltage (top) and representation of electrostatic and mechanical forces as a function of the displacement of the mobile electrode and V_0 (bottom). For a color version of figure, see www.iste.co.uk/basset/keh.zip*

First of all, the designer should locate the stoppers at the dynamic pull-in limit ($x_{unstable}$). However, it is not simple because $x_{unstable}$ evolves with the bias voltage: for high V_0, $x_{unstable}$ is close to $x_{pi} = d_0/3$, while it drifts toward d_0 when V_0 decreases, as illustrated in Figure 7.8. So, there is a trade-off between high biases and large capacitance variation. In addition, a large capacitance variation can often only be obtained at a price of very high sensitivity of the system performances with regard to the amplitude of the external vibrations. Indeed, a large C_{max}/C_{min} can only result from a large ratio between the transducer's gap when the transducer electrodes are close (the gap is $d_0 - X_0$, where X_0 is the amplitude of the mass vibration and d_0 is the gap at the equilibrium), and when the electrodes are far (the gap is $d_0 + X_0$). Hence, if d_0 and X_0 are comparable, a very small variation in the vibration magnitude results in a very large variation of C_{max}/C_{min} ratio [GUI 12].

7.2.2.2. *From monostable to multistable systems*

For the IPGC architecture of Figure 7.5(b) biased by a fixed voltage V_0, the potential function of the system can be expressed as:

$$W_p(x) = \frac{1}{2}kx^2 - \frac{1}{2}C_t(x){V_0}^2 \qquad [7.6]$$

where $C_t(x)$ is given by equation [7.3], k is the stiffness of the resonator. It is important to understand that this is not the potential energy of the system, but a function whose gradient gives the total force applied to the mobile mass (see the discussion about the calculation of the force for a transducer biased by an external voltage source in section 2.3.2). Figure 7.9 shows the evolution of this function versus the mass position for various bias voltages V_0. It can be observed that when the transducer's force is smaller than the spring restoring force, the system has only one stable position at $x = 0$. When V_0 increases, the potential function tends to flatten and two maxima appear corresponding to the unstable positions defined by $x_{\max}$ in the previous section.

We would like to emphasize that the potential function can only be used for qualitative description of the harvester operation. Indeed, equation [7.6] does not account for the dissipative part of the transducer force, which appears when the transducer's bias is not fixed, but varies synchronously with the mobile mass motion. A more accurate description of the system operation requires the use of the analytical tools described in Chapters 5 and 6.

7.2.3. *Capacitive transducers with overlapping electrode motion*

In e-KEH with overlapping electrode motion, the capacitance varies linearly with the mass displacement (assuming the electrode's shape is rectangular), which makes them suitable when the external acceleration amplitude alternates between small and large values. Another great advantage compared to the gap-closing type is that the air friction losses are much smaller since the Couette damping is now negligible.

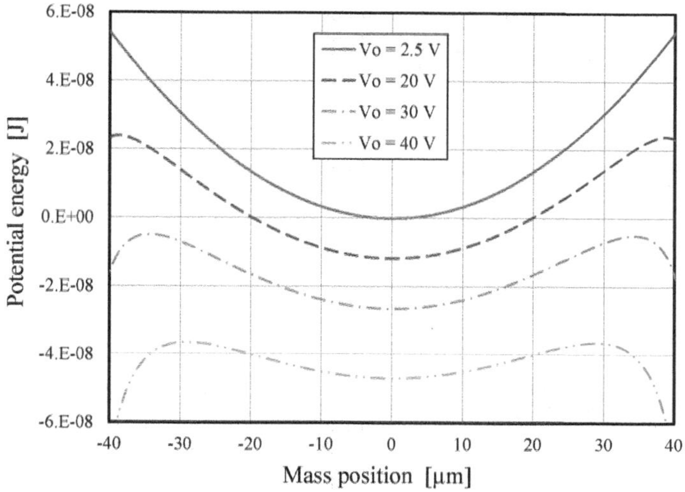

Figure 7.9. *Shape of the potential energy of the IPGC e-KEH in Figure 7.6 with respect to the mass position and for increasing bias voltage [COT 13]. For a color version of figure, see www.iste.co.uk/basset/keh.zip*

7.2.3.1. *In-plane overlap-plate (IPOP) transducers*

IPOP e-KEHs are fabricated by processing a first set of electrodes on one substrate and a second set on another substrate, both substrates being bonded eventually as shown in Figure 7.10. Figure 7.11 shows an example of such a device and its associated process of fabrication in a bulk silicon-glass technology [LIM 06, BAS 09].

With the IPOP configuration, the frequency of the capacitance variation cycle can be much higher than the mechanical frequency: if the mass displacement is higher than the width of one electrode's finger, a single one-way traveling of the mobile electrode can create an overlapping motion of several electrode fingers, which increases the number of energy conversion cycles and then the harvested power. For comparison, this frequency-up amplification factor between the energy conversion cycle and the mechanical oscillation does not exist for the OPGC in Figure 7.5(a) (i.e. is equal to 1) and is equal to 2 for the IPGC of Figure 7.5(b).

The IPOP architecture can be used for the harvesting of vibrations along two axis (two-dimensional KEHs). Examples are given in [PAR 09] (see Figure 7.11(c)) and [BAR 10].

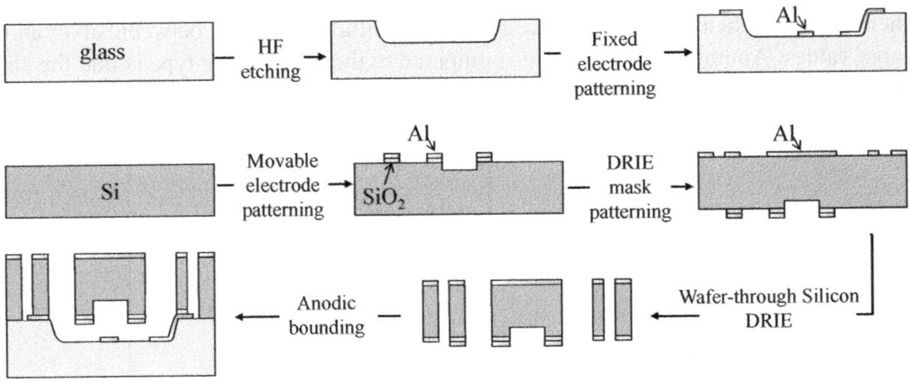

Figure 7.10. *Example of a fabrication process for IPOP MEMS e-KEH*

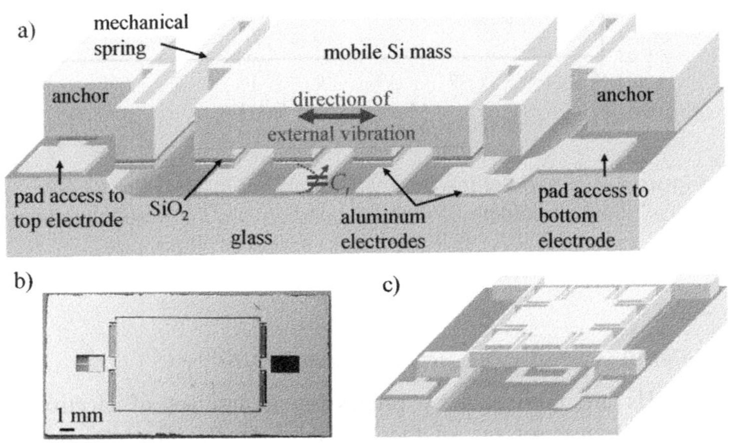

Figure 7.11. *a) Schematic and b) top-view photograph of a 1D in-plane overlap-plate (IPOP) MEMS e-KEH [BAS 09]. c) Schematic of an IPOP 2D MEMS e-KEH [PAR 09]*

The two main challenges with IPOP geometries are to limit the parasitic capacitances and to avoid the pull-in effect. The parasitic capacitance can be substantially reduced by partilally etching the silicon substrate between the fingers of the movable electrodes. Finite element modeling (FEM) simulations show that more than 60% of the parasitic capacitance can be suppressed by performing a partial etching of 20 μm for the devices shown in Figure 7.11 [PAR 07]. To avoid pull-in while increasing the bias voltage, a solution is to include microball bearings to support the mass displacement [NAR 09].

7.3. e-KEHs with an electret layer

The bias voltage, requested for the electrostatic energy conversion, can be generated by an embedded battery or any kind of external power sources, but a more convenient method is to use an electret layer on (at least) one of the transducer's electrodes, see detailed description and analysis of the capacitive transducer with an electret layer in section. 2.3.3. Typical materials for electret used in MEMS are silicon dioxide, teflon, parylene or cytop. The electret is obtained by ionizing the surrounding air in order to electrically charge these materials.

The planar geometry in IPOP e-KEHs makes them very suitable for charging by corona discharge. A bias of a few hundred volts can be reached. As for the devices in Figure 7.11, the supporting material of the electret layer can be pre-etched to reduce the minimal capacitance of the transducer [BOI 10]. Electret charging can also be obtained by air ionization using soft X-rays or vacuum UV [SUZ 11]. These techniques are particularly suitable for charging vertical electret on interdigited-combs, although for a large gap between the fixed and movable fingers, a few dozen volts can be reached with a corona setup [LU 15].

Being able to measure the bias voltage of an e-KEH having an electret layer is important to evaluate the charging process and the evolution of this voltage with time. Indeed, the internal bias tends to quickly decrease the days or weeks following the charging before remaining (almost) constant, depending on the material use for the electret, the conditions of the charging experiment and the device's use and storage (like the humidity or the temperature). The simplest way is to use an electrostatic voltmeter. However, these apparatus are expensive and, in case of a non-contact tool, can hardly measure the bias of vertical electrets in comb-drives. An indirect measurement of the built-in electret voltage can be obtained using the half-wave rectifier circuit presented and analyzed in Chapters 8 and 11. If the capacitance ratio C_{max}/C_{min} is known, the built-in voltage V_{el} can be determined from the saturation voltage measurable at the output of the network through the formula [11.4].

8

Basic Conditioning Circuits for Capacitive Kinetic Energy Harvesters

8.1. Introduction

Any sensor converting a physical quantity into electricity needs a minimal electrical conditioning: an appropriate biasing, a readout circuit with an optimal input impedance, etc. However, in many cases, the conditioning is more complex: for example, resonant sensors need electrical oscillating circuits, zero displacement accelerometers require a sophisticated pulse width modulation or sigma-delta feedback loops, etc. The capacitive transducer, when used for electromechanical energy conversion, also requires a conditioning, whose purpose is to implement a cyclic charge flow synchronized with the variation of the transducer capacitance, as it was discussed in Chapter 2. This chapter presents the basic information about the conditioning circuits, and studies in-depth the most simple conditioning circuit, the *continuous conditioning circuit*.

The case of an energy converter is different from a sensor whose purpose is a measurement of a physical quantity. Whereas a measurement should introduce the lowest possible perturbation into the system, an energy conversion aims to maximize possible energy fluxes from one physical domain into another. As a result, the dynamics of both mechanical and electrical parts is impacted. For this reason, a bi-directional coupling between the two domains is present which cannot be neglected during the analysis and design of the conditioning electronics.

As was shown in Chapter 2, a capacitive transducer needs a dynamic biasing in order to generate electricity. For a transducer whose capacitance varies periodically following a time law $C_t(t)$, the energy conversion process is completely defined by the charge-voltage cycle Γ. Since the voltage, charge and capacitance are bounded by the equation $Q = CV$, if $C_t(t)$ is fixed, the energy conversion is defined by the voltage waveform applied to the transducer. A conditioning circuit for an electrostatic transducer can be seen as an electrical dipole, which is connected to the electrical terminals of the transducer, as presented in Figure 1.3 and which generates the voltage required for implementation of the QV cycle.

Another important role of the conditioning electronics is the interface with the load. In general, it means that the harvested energy must be transferred into some energy reservoir (a capacitor and a battery) having a low and stabilized voltage. Both the energy sensor conditioning and interface with load should be implemented at a minimum energy cost. This last point is paramount for capacitive energy harvesters. Indeed, in the numerical example of section 2.5, we showed that the rate of energy converted by a capacitive transducer implemented in a microelectromechanical systems (MEMS) technology is, at best, a few dozen microwatts. Hence, in the best cases, only a few microwatts are available for operation of the conditioning circuit.

As a result, the two roles of the conditioning circuits combined with the ultra low-power requirement create strong challenges for its design.

The synchronization of the charge flow on the capacitive transducer with the motion of the mobile electrode (e.g. as required by the constant voltage QV cycle) is also very challenging. Existing MEMS devices for electrostatic vibration energy conversion contain only one capacitive transducer used for the energy conversion, and it is impossible to measure the instantaneous position of the mobile mass.

The conditioning circuits are discussed in Chapters 8–11. This chapter begins by an overview of conditioning circuits, and presents few widely used elementary topologies. Their shortcomings motivate the implementation of more complex architectures. Chapter 9 presents some basics about the transfer of energy in capacitive networks, and presents the family of conditioning circuits implementing triangular charge-voltage cycles. Chapters 10 and 11 present the most advanced conditioning circuits known to date, which implement rectangular charge-voltage cycle.

8.2. Overview of conditioning circuit for capacitive kinetic energy harvesting

Since the last two decades, several architectures of conditioning circuits for e-KEHs were invented. We propose a classification of the conditioning circuits on the basis of the shape of the achieved charge-voltage cycles. Figure 8.1 presents all known basic QV cycles plotted in the same axes, under hypothesis that the maximum voltage V_t applied to the variable capacitor is the same for all cycles (V_{max}). We chose such a basis for the comparison, because in practice, any technology of electronic circuits always imposes a maximum allowed voltage. We distinguish three families of conditioning circuits implementing three types of QV cycles:

1) teardrop, oval or egg-shaped QV cycle (an example is plotted with dotted lines);

2) triangular QV cycle (OAB and OCB);

3) rectangular QV cycle (DFBE).

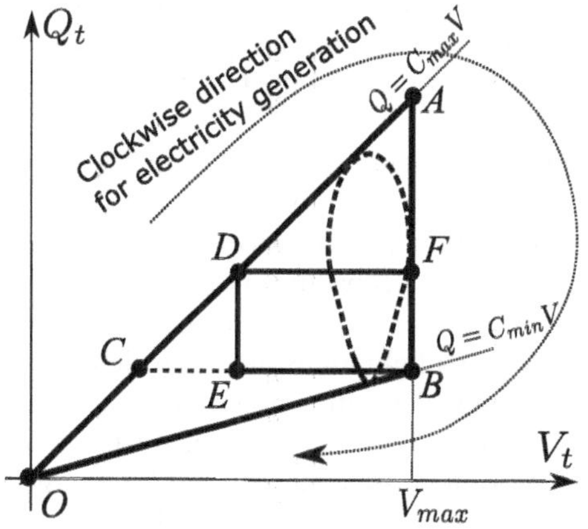

Figure 8.1. *Diagram showing all charge-voltage cycles which have been used to date for conditioning of capacitive transducers in KEHs: OCB (constant-charge), OAB (constant-voltage), DFBE (rectangular) and "tear drop" cycle corresponding to the continuous conditioning circuit or KEH with electret layer. All cycles are drawn for the same extreme values of the variable capacitance* (C_{min}, C_{max}) *and for the same maximum voltage* V_{max}

This chapter is devoted to the first family of circuit, which is called "continuous" or "primitive" conditioning circuit. The two other families are discussed in Chapters 9–11.

When conditioning circuits are studied, the variation of the transducer capacitance is supposed to be defined, and fully characterized by a function of time $C_t(t)$. It is usually assumed that $C_t(t)$ is periodic with period T and having only one local maximum and minimum (C_{max} and C_{min}) over a period. This is only a working hypothesis allowing us to limit the complexity of the analysis. In reality, because of the electromechanical coupling, $C_t(t)$ also depends on the electrical operation of the conditioning circuit. However, assuming a predetermined $C_t(t)$ is a necessary step in the study of the harvester as a whole.

8.3. Continuous conditioning circuit: generalities

The simplest conditioning circuit demonstrating a generation of electrical power out of variation of a capacitance is the "continuous conditioning circuit". Its topology is presented in Figure 8.2. The circuit is composed of the variable capacitor C_t (the transducer), a resistive load R_L and an initially charged large reservoir capacitor C_{res}, or simply a DC voltage source, when only a laboratory test is aimed. The name of the circuit comes from the fact that all voltages and currents of the circuit are continuous functions of time; this is not the case of more sophisticated circuits which use switches or diodes and which will be considered in the Chapters 9–11.

The continuous conditioning circuit presented in Figure 8.2 was first discussed in [STE 03, MIT 08]. It was proposed as an electric interface between a resistive load and the capacitive transducer provided with an electret layer (see Chapter 2) [STE 03], and later it was used with a passive capacitive transducer [MIT 08, BAS 14] biased by a voltage source or by a fixed precharged capacitor. In spite of the different natures of transducer devices, these configurations are equivalent: as we showed in section 2.3.3, a DC voltage source in series with a passive capacitive transducer is exactly an electrical model of a transducer biased by an electret layer. Without any impact on the model, the DC voltage source may be replaced by a large precharged capacitor. Because of the simplicity of its implementation, this circuit is also very often used in laboratory tests and characterization of the capacitive transducers intended for the energy conversion. In this section, we propose a brief analysis of this circuit and a discussion of its operation. We also highlight its essential shortcomings: that it is important to understand the motivation of investigation of more sophisticated solutions which are presented in the Chapters 9–11.

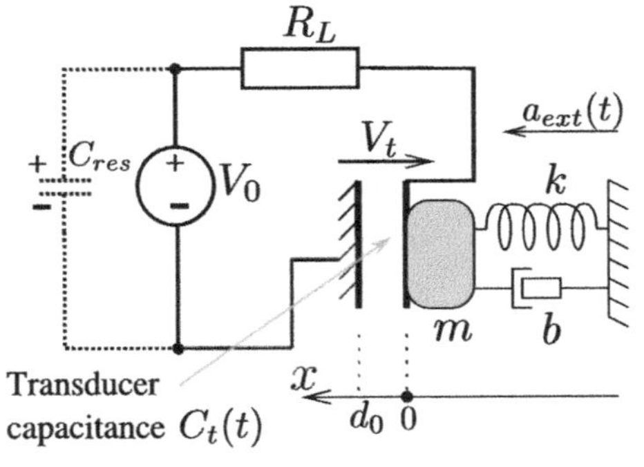

Figure 8.2. *Schematic of continuous conditioning circuit. In this study, the continuous circuit represents all circuit configurations which employ a capacitive transducer, resistive load and some internal or external voltage to provide the initial bias on the transducer*

8.3.1. *Qualitative discussion on operation of the circuit*

In this section, we briefly explain the circuit operation on the example of a circuit where the biasing is provided by a large fixed capacitor C_{res} initially charged to a voltage V_0, and where the transducer capacitance varies according to some known periodic time law $C_t(t)$ (Figure 8.2(a)). We first consider the circuit operation in an extreme case, when R_L is close to zero [BAS 14]. In this case, the voltage on C_t can be considered constant in time since we can neglect the voltage on R_L comparing to V_0. The transducer charge varies as $V_0 C_t(t)$, and hence there is a charge flow between C_t and C_{res} corresponding to a current $i(t)$ equal to:

$$i(t) = V_0 \frac{\mathrm{d}C_t}{\mathrm{d}t} \qquad [8.1]$$

It can be seen that this current dissipates power on R_L, equal to:

$$P_{RL} = i^2(t) R_L. \qquad [8.2]$$

After each period of variation of $C_t(t)$, the circuit returns to the same electrical state, since the total electrical charge of C_t and C_{res} is constant, so is their total energy. We should conclude that the energy dissipated on the resistance corresponds to the conversion of energy from mechanical to electrical domain.

The power P_{RL} is proportional to the load resistance: it increases as R_L increases. However, for large R_L, the voltage on R_L cannot be neglected, and the equation [8.1] is not valid anymore.

Let us consider the opposite extreme value of the load resistance: a very large (infinite). If the power on the load resistance is expressed through the voltage on the load resistance V_L as:

$$P_{RL} = V_L^2(t)/R_L, \qquad [8.3]$$

we can show that the power tends to zero as $R_L \to \infty$. To prove it, it is enough to show that V_L has an upper bound. Indeed, since the current tends toward zero, the transducer keeps its charge constant. The value of this charge depends on the initial condition. Let it be Q_0, in this case, the voltage on C_t is $Q_0/C_t(t)$, and the voltage on the resistor is equal to $V_0 - Q_0/C_t(t)$. If C_{min} is different from zero, this time function has an upper bound.

So, at zero and infinity values of the load resistance, the power is zero, but the power is non-zero for finite values of R_L. Hence, there is an optimal value of the load resistance, for which the converted power is maximum.

In sections 8.4 and 8.5, we propose a more detailed analysis of the circuit.

8.3.2. *Analytical model in the electrical domain*

This section presents the mathematical model formally describing the operation of the conditioning circuit.

The governing equations describing the electrical behavior of the simple conditioning circuit are given by the Kirchhoff voltage law and the element equations:

$$R_{\mathrm{L}}\frac{dQ_t}{dt}+\frac{Q_t}{C_t(t)}=V_0, \qquad [8.4]$$

where Q_t is the instantaneous charge on the transducer capacitor and $C_t(t)$ is the time evolution law of the transducer capacitance. The electrical equation of the transducer is simply

$$V_t(t)=Q_t(t)/C_t(t), \qquad [8.5]$$

where V_t is the voltage on the transducer.

The instantaneous power converted by the transducer is dissipated on the load resistance, and can easily be calculated with equation [8.2], where the current is given by:

$$i=\frac{\mathrm{d}Q_t}{\mathrm{d}t}. \qquad [8.6]$$

The average power converted by the transducer is given by the average of [8.2] on one period of $C_t(t)$ variation.

In order to calculate the average converted power, the closed expression of $Q_t(t)$ should be found, and for that the differential equation [8.4] should be solved. The work of Galayko [GAL 13a] addressed the resolution of this equation. It was concluded that even for simple geometries of the transducer, this equation cannot be solved in closed form, and the analytical solution for $Q_t(t)$ can only be as an infinite (Fourier) series, in which, however, the terms above third or fourth can be neglected in the most practical cases. The coefficients of the series are expressed through Bessel functions. For this reason, a comprehensive analytical expression relating the converted power to the parameters of the system is difficult to establish.

8.4. Practical study of continuous conditioning circuits

In order to practically study the operation of the continuous conditioning circuit, we propose to use a Spice simulator to numerically solve the circuit equation. Note that any other tool solving differential equations can be used (e.g. Matlab, Scilab, etc.).

In this study, the mobile electrode of the capacitive transducer is assumed to move according to the sinusoidal law:

$$x(t) = X_0 \cos(\omega t). \qquad [8.7]$$

The transducer capacitance-to-displacement characteristic depends on the geometry of the transducer. In this study, we consider two kinds of transducer: the gap-closing geometry and the area overlap geometry, (see section 2.1). The corresponding $C_t(x)$ functions are the following:

– for the gap-closing transducer:

$$C_t(x) = \varepsilon_0 \frac{S}{d - x}, \qquad [8.8]$$

– for the area overlap transducer:

$$C_t(x) = C_0 + \alpha x, \qquad [8.9]$$

where $C_0 = \varepsilon_0 \frac{lw}{d}$, $\alpha = \varepsilon_0 \frac{w}{d}$ (see equations [2.6] and [2.5]).

We note that the law of the $C_t(t)$ variation is important only for the continuous conditioning circuit; for the circuits which will be considered in the Chapters 9–11, only the extreme values on the period C_{min} and C_{max} matter.

8.4.1. *Gap-closing transducer*

In the case of the gap-closing geometry, the circuit equation [8.4], when combined with [8.7] and [8.8], is given by:

$$\frac{\mathrm{d}Q_t}{\mathrm{d}t} + \frac{d}{R_L \varepsilon_0 S} Q_t \left(1 - \frac{X_0}{d} \cos \omega t\right) = \frac{V_0}{R_L}. \qquad [8.10]$$

In order to study the influence of the parameters of the circuit on the behavior, we propose to introduce a normalization of the equation variables. As we will show, this will reduce the number of the system parameters to a minimal set including only independent parameters. As a result, the space of parameters will reduce, and we can focus the study only on the meaningful parameters. In order to obtain a normalized equation, we propose the following steps:

– we introduce a normalized dimensionless time $\tau = \omega t$, so that the period of the capacitance variation is now 1. In this case, the derivation operator $\frac{\mathrm{d}}{\mathrm{d}t}$ is equal to $\omega\frac{\mathrm{d}}{\mathrm{d}\tau}$, and equation [8.10] is written as:

$$\frac{\mathrm{d}Q_t}{\mathrm{d}\tau} + \frac{d}{R_L\varepsilon_0 S\omega}Q_t\left(1 - \frac{X_0}{d}\cos\tau\right) = \frac{V_0}{R_L\omega}. \qquad [8.11]$$

– it is obvious that the last equation has only four meaningful parameters (coefficients), comparing to six parameters of the initial equation. We propose to name them as:

- the normalized amplitude of mobile electrode motion, y_0:

$$y_0 = \frac{X_0}{d}, \qquad [8.12]$$

- the normalized RC time constant of the circuit. The readers may notice that $\varepsilon_0 S/d$ is the value of the transducer capacitance at $x = 0$, its product with R_L is related to the relaxation time of the circuit (although, since the system is nonlinear, the relaxation time is not a constant and depends on the trajectory of the mobile electrode motion). Because of the presence of ω, the coefficient

$$\rho = \frac{d}{R_L\varepsilon_0 S\omega} \qquad [8.13]$$

can be named “normalized relaxation time” of the circuit,

- the coefficient

$$I_0 = \frac{V_0}{R_L\omega} \qquad [8.14]$$

has the dimension of a normalized current (derivative of the charge over normalized time), and represents the biasing of the circuit;

We obtain the following equation:

$$\frac{\mathrm{d}Q_t}{\mathrm{d}\tau} + \rho Q_t\left(1 - y_0\cos\tau\right) = I_0. \qquad [8.15]$$

– observing equation [8.15], we note that the solution for $Q_\tau(\tau)$ is proportional to I_0, since the left part of the equation is linear with regard to Q_t. As a result, we can eliminate the parameter I_0 by introducing the normalized dimensionless charge $\theta = q/I_0$. We obtain the final normalized equation containing only two independent parameters, ρ and y_0:

$$\frac{\mathrm{d}\theta}{\mathrm{d}\tau} + \rho\theta\,(1 - y_0 \cos\tau) = 1. \qquad [8.16]$$

The advantage in the use of a normalized equation is that instead of having six independent parameters in the original equation [8.10] (d, X_0, ω, S, V_0, R_L), we now have only two. So, the parametric study of the circuit behavior is much more simple. We propose to fix four parameters of the system (for instance, d, ω, S, V_0), and to vary two others (X_0 and R_L) which impact independently the normalized parameters ρ and y_0.

Table 8.1 provides the numerical values of the parameters for which we propose a study of the system. We propose to select three values of y_0 corresponding to three amplitudes of the mobile mass vibration X_0, and for each amplitude we sweep the value of the parameter ρ (the load resistance R_L).

Physical parameters						
	Fixed parameters				Variable parameters	
Parameter name	d (m)	S (m^2)	ω (rad·s^{-1})	V_0 (V)	X_0 (m)	R_L, (Ω)
Parameter value	$50\cdot10^{-6}$	$1\cdot10^{-4}$	$2\cdot\pi\cdot100$	10	$(30,40,45)\cdot10^{-6}$	$(1...100)\cdot10^{6}$
Normalized parameters						
Parameter name	y_0			ρ		
Parameter value	0.6, 0.8, 0.9			89.92...0.8992		

Table 8.1. *Values of parameters used for simulation of continuous conditioning circuit with gap-closing capacitive transducer*

Figure 8.3 presents the plot relating the average power generated on the load resistance in the steady-state mode (after the transient process) to the value of the load resistance. This plot is given for three amplitudes of the mobile mass vibrations. Similar curves are obtained for the power generated on a load resistance by an ideal voltage source having an internal resistance R_s (see the discussion in Chapter 1). In this case, the power is maximum when the load resistance is equal to the internal resistance of the source. This allows us to consider the load resistance value at which the power on the plot of Figure 8.3 is at its maximum as the equivalent internal resistance of the power source represented by the transducer and the biasing source. Note that the internal resistance of such a source depends on the amplitude of the mobile electrode motion, as shows the plot.

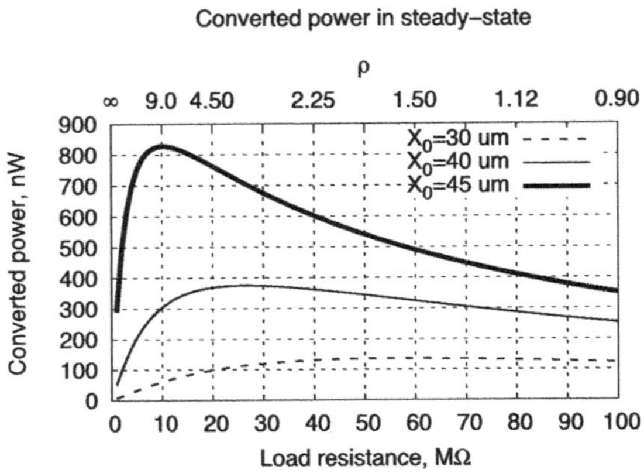

Figure 8.3. *Average power converted in a steady-state by the continuous conditioning circuit used with a gap-closing transducer against the load resistance, for three different amplitudes of the mobile electrode motion*

Figure 8.4 presents three families of QV cycles in steady-state mode (after the end of the transient process), corresponding to three different amplitudes of the mass displacement X_0. Each plot contains several QV cycles obtained for different values of the load resistance. The QV cycles drawn with a thick plain line correspond to the values of the load resistance at which the converted power is maximum (see Figure 8.3): this QV cycle has the largest area over all cycles of the family.

Let us have a look on the evolution in the shape of the QV cycle as the load resistance increases. For low R_L, the QV cycles are more "vertical", and as $R_L \to 0$, the cycle is degenerated to a vertical line (meaning that there is no voltage across the load resistance). On the contrary, for large R_L, the QV cycle is more horizontal, and in the limit case when $R_L \to \infty$, the QV cycle is a horizontal line (meaning that there is no current flowing through the load resistance). For these two extreme cases, the energy converted by a cycle is zero (see discussion in section 8.3.1). The QV cycle has a "tear-drop" shape, which, as we discussed before, cannot be expressed by any analytical function written in closed form.

The plots of Figure 8.5 present the time evolution of the transducer capacitance, the transducer voltage and current at the configuration where $X_0 = 45\ \mu m$, $R_L = 10\ M\Omega$. The transient process lasting for less than one period is observable on these plots. The nonlinear nature of the system can clearly be seen from these curves.

In section 8.5.1, we provide the Spice netlist written for the Eldo simulator of Mentor Graphics, which was used for the generation of the discussed plots.

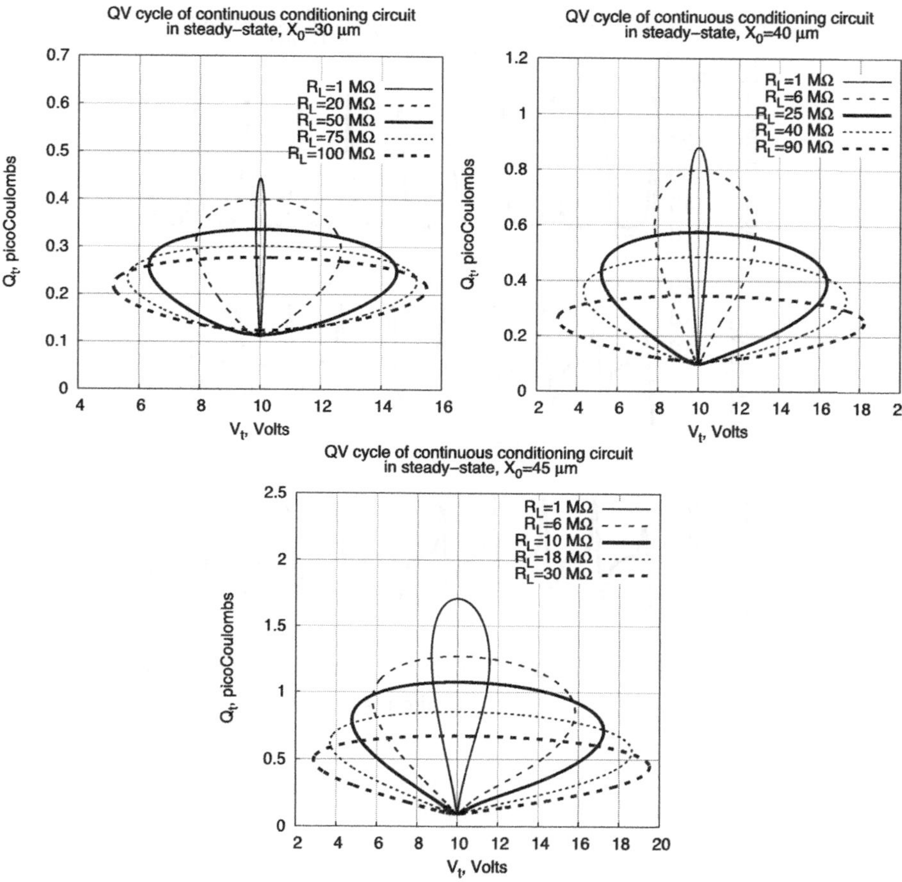

Figure 8.4. *Three families of QV cycles corresponding to steady-state operation of the continuous conditioning circuit used with a gap-closing transducer, each family is plotted for a value of the amplitude of mobile electrode motion. The QV cycles on each plot correspond to different load resistances. The optimal QV cycle (those maximizing the converted power and hence having a maximum area) is plotted with thick solid line*

8.4.2. *Area overlap transducer*

The analysis of the circuit behaviors with area overlap transducers follows the same algorithm as with the gap-closing transducer. Hence, we briefly present the main steps of the analysis, and we only emphasize the differences from the case of the gap-closing geometry.

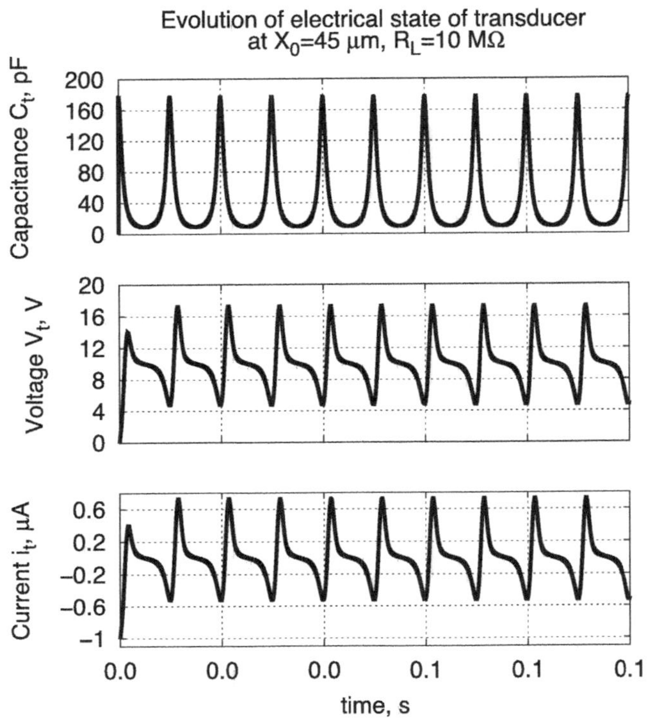

Figure 8.5. *Time evolution of the electrical quantities in the continuous conditioning circuit used with a gap-closing transducer, obtained by simulation. The transient and the steady-state behavior are clearly observed*

With a transducer having the area overlap geometry, the circuit equation [8.4], when combined with equations [8.7] and [8.9], is given by:

$$\frac{\mathrm{d}Q_t}{\mathrm{d}t} + \frac{1}{R_L} Q_t \frac{1}{C_0 + \alpha X_0 \cos \omega t} = \frac{V_0}{R_L}. \qquad [8.17]$$

We normalize the equation by doing the following substitution:

$$t = \tau/\omega, \; Q_t = \theta \frac{\omega R_L}{V_0}, \qquad [8.18]$$

and by introducing the normalized parameters:

$$\rho = \frac{1}{C_0 R_L \omega}, \; y_0 = \frac{\alpha X_0}{C_0}. \qquad [8.19]$$

We obtain the following normalized equation having two independent parameters:

$$\frac{d\theta}{d\tau} + \rho\theta \frac{1}{1 + y_0 \cos\tau} = 1, \qquad [8.20]$$

As in the case of the gap closing transducer, the solution of this equation cannot be expressed in closed form, so we use a Spice simulation in order to study the behavior of the circuit (see section 8.5.1). We fix all parameters of the original equation [8.17], except the mobile mass displacement amplitude X_0 and the load resistance R_L. The numerical values of the parameters used in the simulation are given in Table 8.2. We run the simulation for two values of amplitude ($30\mu\, m$ and $50\mu\, m$), for several values of the load resistance.

<table>
<tr><td colspan="7">Physical parameters</td></tr>
<tr><td></td><td colspan="4">Fixed parameters</td><td colspan="2">Variable parameters</td></tr>
<tr><td>Parameter name</td><td>C_0 (F)</td><td>α (F·m^{-1})</td><td>ω (rad·s^{-1})</td><td>V_0 (V)</td><td>X_0 (m)</td><td>R_L (Ω)</td></tr>
<tr><td>Parameter value</td><td>$200\cdot10^{-12}$</td><td>$1\cdot10^{-6}$</td><td>$2\cdot\pi\cdot100$</td><td>10</td><td>(30 and 50)·10^{-6}</td><td>(1...100)·10^{6}</td></tr>
<tr><td colspan="7">Normalized parameters</td></tr>
<tr><td>Parameter name</td><td colspan="4">y_0</td><td colspan="2">ρ</td></tr>
<tr><td>Parameter value</td><td colspan="4">0.15 and 0.25</td><td colspan="2">7.9577...0.0796</td></tr>
</table>

Table 8.2. *Values of parameters used for simulation of continuous conditioning circuit with area overlap capacitive transducer*

Figure 8.6 presents the average power converted in the steady-state mode. The aspect of the plot is very similar with the same characteristic for the gap closing transducer (Figure 8.3). An important difference with the case of the gap closing transducer is that the optimal value of the load resistance maximizing the converted power is the same for different amplitudes.

Figure 8.7 presents the family of QV cycles plot for a given amplitude X_0, and for several values of load resistances. The solid thick lines correspond to cycles with the largest converted power. We can see that the shape of the QV cycles is different from the case of gap closing transducer: the cycles resemble ovals, however, a careful observation of the cycles corresponding to the maximum converted energy (thick solid lines) shows that the shape is rather close to that of an egg. Again, no analytical function expresses precisely this shape.

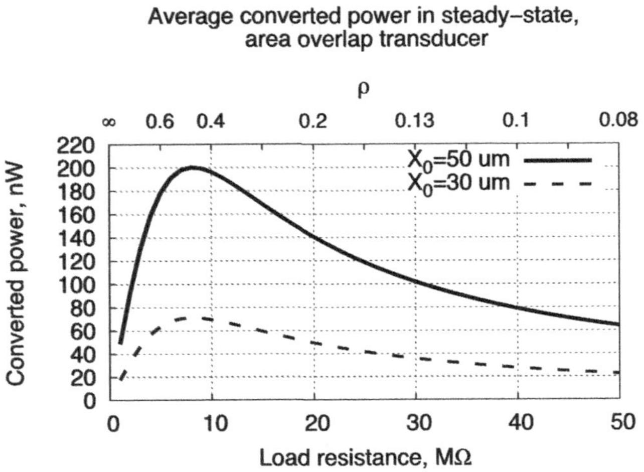

Figure 8.6. *Average power converted in a steady-state by the continuous conditioning circuit used with a area overlap transducer against the load resistance, for three different amplitudes of the mobile electrode motion*

Figure 8.8 presents the evolution in time of the electrical quantities associated with the transducer plotted for $X_0 = 50\ \mu m$ and $R_L = 8\ M\Omega$. The aspects of the voltage and current waveforms are quite different from the similar curves plotted for the gap closing transducer.

8.4.3. *Simple conditioning circuit with diode rectifiers*

One of that drawbacks of continuous conditioning circuit is an AC voltage generated on the load. This can be seen by considering the current flowing through the transducer in Figures 8.5 and 8.8. An AC-DC conversion is required for the load supply. Two circuits presented in Figure 8.9 are used for rectification of the output voltage. These topologies have been widely used with capacitive transducers biased by an electret layer; they represent very basic networks known from the power electronics.

The circuits presented in Figures 8.9(a) and (b) are half-wave and full-wave diode rectifiers. They make use of the variation of the transducer capacitance in order to accumulate electrical charges on a large reservoir capacitance C_{res}, which then may be used as a DC supply source for a resistive load.

Although these topologies are naturally derived from the continuous conditioning circuit, their operation is very different. They key figures of performance of the circuits based on the diode rectifiers (the power yield, the optimal mode of operation) can be obtained by a simple analysis in the electrical domain. These circuits implement a *rectangular* QV cycle, and for this reason, we will discuss them in Chapter 11.

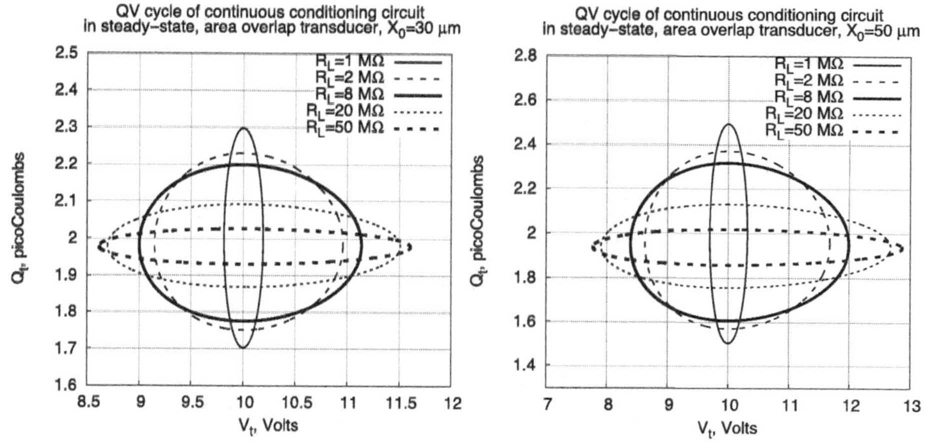

Figure 8.7. *Two families of QV cycles corresponding to steady-state operation of the continuous conditioning circuit used with an area overlap transducer, each family is plotted for a value of the amplitude of mobile electrode motion. The QV cycles on each plot correspond to different load resistances. The optimal QV cycle (those maximizing the converted power and hence having a maximum area) is plotted with thick solid line*

8.5. Shortcomings of the elementary conditioning circuits: auto-increasing of the biasing

The main advantage of the continuous conditioning circuit is its simplicity of implementation. Unfortunately, the drawbacks are numerous, and this is the main motivation for study of alternative but more complex solutions. In this conclusion, we summarize the drawbacks of this conditioning circuit:

– the voltage on the load resistance is AC; it can be seen from the current flowing through the transducer in Figures 8.5 and 8.8. An AC-DC conversion is required; in the case of the electret transducer, this is done by conventional rectifying networks;

– the output voltage may be of the same order of magnitude as the bias voltage. In the case of the electret layer, the bias voltage may be of several tens or even hundreds of volts, and a downscalling of the voltage is necessary. This is related to generally very large internal impedance (resistance) of the energy source provided by the continuous conditioning circuit (see discussion in section 8.4.1). An active impedance matching is required in order to optimally supply an eventually low impedance load;

– however, the main shortcoming of the continuous conditioning circuit is the impossibility to increase its initial biasing. Practical studies highlight that up to dozens of volts of bias may be needed for optimal operation [BAS 14]. However, the initial charge can only be obtained by a generally low-voltage external source (battery, piezoelectric or solar cell [FLO 11], RF waves, etc.). In any case, the initial voltage generated in this way will probably be low (few volts). A low bias voltage means a low output power, since the latter is proportional to the square of the voltage. A similar problem is when the electret layer is weakly biased, for example, because of the depolarization due to the aging;

– if the bias voltage is generated by a large precharged capacitor, the leakage reduces its charges with time, and the circuit becomes inactive.

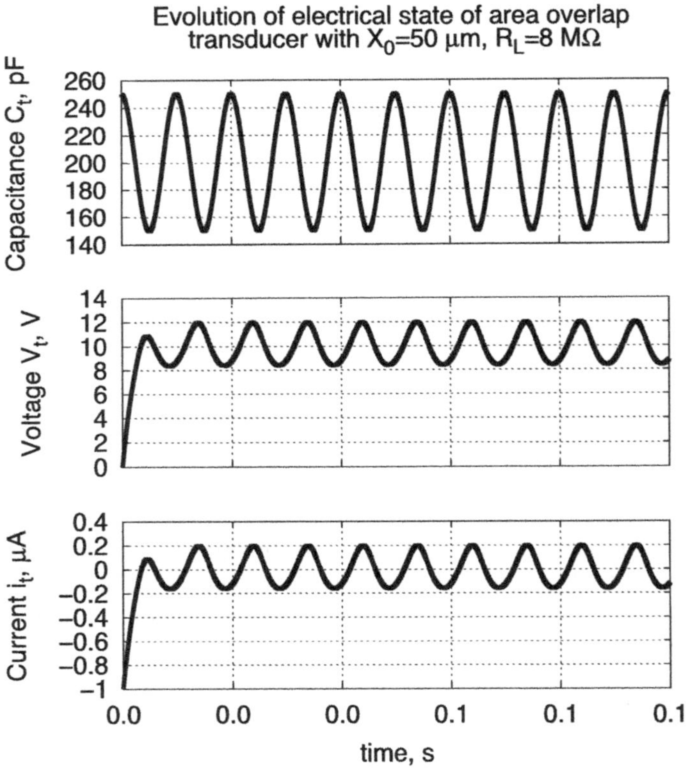

Figure 8.8. *Time evolution of the electrical quantities in the continuous conditioning circuit used with an area overlap transducer, obtained by simulation. The transient and the steady-state behavior are clearly observed*

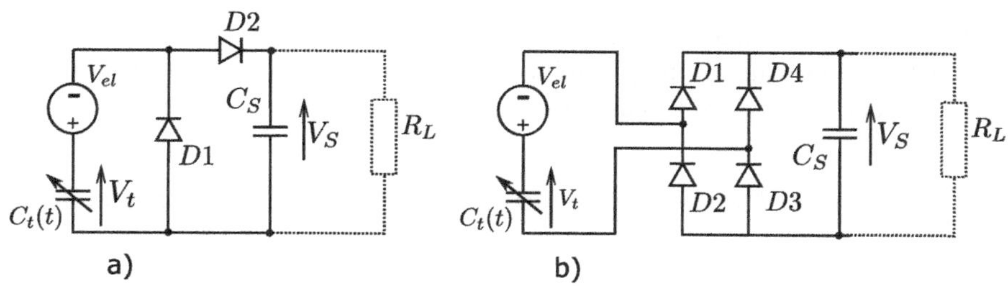

Figure 8.9. *Two circuits based on diode voltage rectifiers used with a transducer biased by electret layer. a) Half-wave rectifier and b) full-wave rectifier*

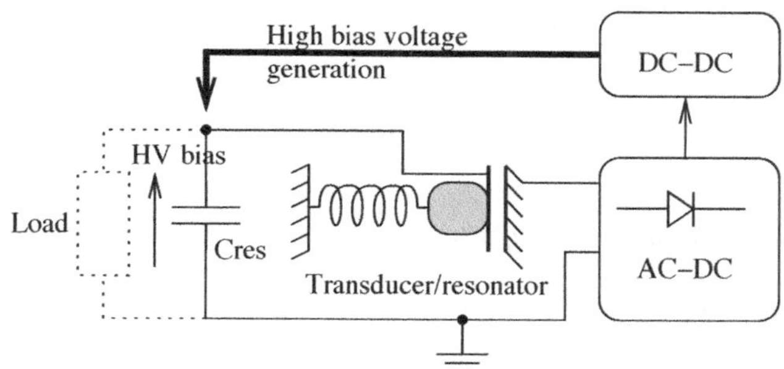

Figure 8.10. *Functional diagram of a conditioning circuit allowing an accumulation of the converted energy on the reservoir capacitor*

The last two points are related to the impossibility of the continuous conditioning circuit to accumulate the energy converted from the mechanical domain, in order to increase its biasing. One of the possible functional diagrams required for a conditioning circuit is presented by the diagram in Figure 8.10: there should be a mechanism allowing a generation of a high voltage by using the energy converted from the mechanical domain. This voltage should then be used for biasing of the conditioning circuit. Since the biasing does not consume energy (under hypothesis of low leakage of capacitors), the circuit will be able to operate at high biasing, while the start-up may be done from a low voltage. Obviously, at the initial stage, the energy converted from the mechanical domain should be used for generation of the high-voltage biasing, and only after the biasing is established, the converted energy can be provided to the load with a high power due to a high-voltage biasing. We call the mode in which the conditioning circuit accumulates its energy in order to increase its biasing "the auto-accumulative" or "self-increasing" mode.

The circuits presented in the Chapters 9–11 are free from the previously mentioned drawback, and provide all the possibility to accumulate the converted energy on a storage capacitor, which is then used for the load supply.

8.5.1. *Appendix: listing of the Eldo netlist used to obtain the presented plots*

```
***** The first line is ignored in spice netlists******

**** The parameter definition *****
******* Fixed parameters******

****For the gap closing transducer*****
*The transducer gap
.param d=50e-6
*The transducer area
.param S=1e-4

***For the area overlap transducer****
*Capacitance at rest
.param C0=200e-12
*The transducer linear coefficient
.param alpha={100e-12/100e-6}


*Permittivity of vacuum
.param eps0=8.85e-12
*Frequency of the mobile plane motion
.param f=100
*Bias voltage
.param V0=10
*Simulation time
.param simtime=0.1

******* Variable parameters******
*Amplitude of the mobile plane motion
.param X0=45e-6
*Load resistance
.param RL=10e6


*The voltage source defining the trajectory of the mobile electrode
Vx x_value 0 sin 0 X0 f 0 0 90

*The voltage controlled voltage source
```

```
*defining the variation of the transducer capacitance
*This is the only line depending on the transducer
*geometry: comment/uncomment when one or another
*transducer is simulated
****For the gap closing transducer
ECt Ct_value 0 value={eps0*S/(d-v(x_value))}
****For the area overlap transducer
***ECt Ct_value 0 value={C0+alpha*v(x_value)}


*The transducer
Ct b1 0 value={v(Ct_value)}

* The bias  voltage source
V0 b2 0 V0

* The load resistance
RL b1 b2 RL

*Transient simulation command,
*with option uic (Use Initial Conditions)
.tran {simtime} {simtime} uic
.step param RL 1e6 100e6 1e6

*Plotting commands
.plot V(b1) V(b2) V(b1,b2)
.plot V(Ct_value)
* to add plotting commands if necessary

*simulator options
.option hmax=1e-4 vntol=3e-9 reltol=0.00005 eps=1e-12 chgtol=5e-12 abstol=1
    e-10 gmin=1e-12 hmin=1e-13
```

Listing 8.1. *Continuous conditioning circuit with gap closing and area overlap transducer: Eldo netlist. Comment/uncomment lines for selection of the transducer model*

9

Circuits Implementing Triangular QV Cycles

This chapter discusses the circuits which implement triangular QV cycles, like the constant voltage QV cycle discussed in Chapter 2. These circuits use active circuits synchronizing the charge flow of the transducers. This chapter is composed of three parts. In section 9.1, we recall some basic notions about the energy transfer in the capacitive circuits. This will help us to understand the difficulties in a practical use of capacitive transducers for the energy conversion. Sections 9.2.1 and 9.2.2 will be devoted to the discussions on the circuits implementing constant-voltage and constant-charge cycles.

9.1. Energy transfer in capacitive circuits

In this section, we discuss the following very basic problem: how to transfer electrical energy from one capacitor to another, while minimizing the energy losses. We consider several configurations which are widely used in conditioning circuits.

9.1.1. *Energy exchange between two fixed capacitors*

Let us assume two identical capacitances C_1 and C_2, such as $C_1 = C_2 = C$, initially have voltages $V_1 = V_0$ and $V_2 = 0$, respectively. Suppose that some kind of energy transfer is required from C_1 to C_2. This problem is very common in power electronics.

In order to transfer energy from C_1 to C_2, it is possible to connect the two capacitors in parallel by a switch SW, whose resistance is modeled by a resistor R (Figure 9.1). It is easy to see, that after a transient process, both capacitors have a voltage $V_0/2$. The energy balance is the following:

– before switching, the total energy of the system is:

$$W_{C1} + W_{C2}\big|_{t<0} = \frac{CV_0^2}{2} \quad [9.1]$$

– a long time after switching, the total energy is:

$$W_{C1} + W_{C2}\big|_{t\to\infty} = 2 \cdot \frac{C(V_0/2)^2}{2} = \frac{1}{2}\frac{CV_0^2}{2}, \quad [9.2]$$

i.e. two times less than the initial energy of the system.

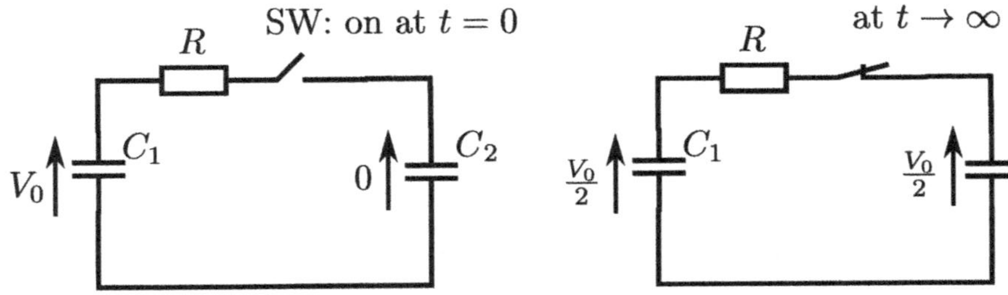

Figure 9.1. *Energy transfer between two capacitors*

Half of the initial energy is dissipated in the wires. What is remarkable is that the dissipated energy does not depend on the value of the resistor: even if $R \to 0$, exactly half of the energy is dissipated. This suggests a fundamental nature of this phenomenon. What happens if in practice we tend R to zero (for example, by using supraconductive materials, etc.) makes for an interesting discussion. In this case, the inductance of the wires is dominant, the circuit oscillates and electromagnetic emission removes the energy from the system, thereby imitating a dissipation. As the equilibrium[1] is reached, the circuit always has half the initial energy, whatever the mechanism of losses.

9.1.2. *Case of a voltage source charging a capacitor*

If a voltage source is connected to an initially discharged capacitor (see Figure 9.2), the situation is very similar to the previous case. The readers may verify that:

1 In the framework of the classical electrodynamic, electromagnetic emission due to alternative current of the oscillations provokes a steady dissipation leading to an equilibrium with $V_{C1} = V_{C2}$. If quantum effects are accounted for, a very small residual oscillation remains, not detectable in practice because of the noise floor.

– the capacitor C is charged to V_0 at the end of the transient process;

– the voltage source spends the energy equal to the total charge Q multiplied by V_0. This is exactly twice the energy acquired by C equal to $QV_0/2$. The difference is dissipated in the connection wires.

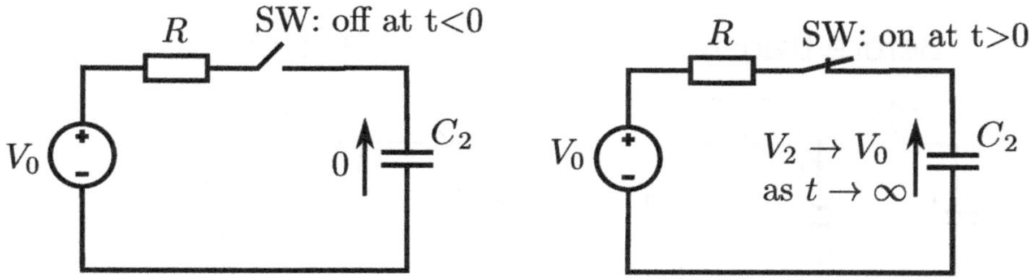

Figure 9.2. *Energy transfer between a voltage source and a capacitor*

In the capacitive energy harvesters, energy transfers with 50% of efficiency are not acceptable. Two options for lossless energy transfer between capacitors are discussed in the next sections.

9.1.3. *Inductive DC-DC converters*

From the basic circuit theory, it is known that the energy of a capacitor can be transferred without losses into an inductor and vice versa. In this way, an inductor can be used as an energy buffer for a lossless transfer of energy between capacitors. This technique is widely used in the power electronics. More information about inductive power converter can be found in any handbooks on the power electronics, for example, [ANG 05] and [ERI 07].

Figure 9.3 presents a network allowing a lossless transfer of energy from one charged capacitor to another initially discharged capacitor. First, C_1 is charged to V_0, the inductor L and the capacitor C_2 are discharged. At $t = 0$, the switch $SW1$ is ON for a time equal to the quarter of period T_1, where $T_1 = 2\pi\sqrt{LC_1}$ is the period of natural oscillations of the network LC_1. At $t = T_1/4$, the voltage in C_1 is zero, and the current in the inductor is equal to $I = \sqrt{C_1/L}V_0$. At this moment, $SW1$ becomes OFF and $SW2$ becomes ON, so that the inductor discharges on C_2. The discharge takes a time equal to $T_2/4$, where $T_2 = 2\pi\sqrt{LC_2}$ is the period of natural oscillations of the network LC_2. At $t = (T_1 + T_2)/4$, the current in the inductor is zero, and all initial energy of C_1 is now in C_2. From the energy balance equation:

$$\frac{C_1 V_0^2}{2} = \frac{C_2 V_1^2}{2}. \qquad [9.3]$$

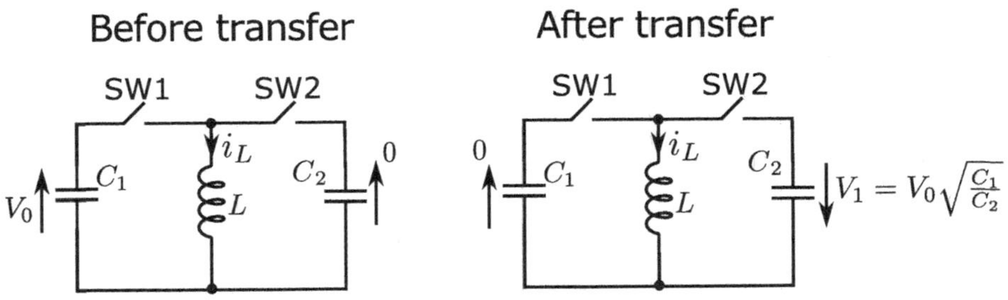

Figure 9.3. *Use of a DC-DC converter for lossless energy transfer between two capacitors*

It is easy to find that $V_1 = \sqrt{C_1/C_2}V_0$. Depending on the ratio C_1/C_2, the voltage V_1 may be greater or less than V_0.

This mechanism can also be used if only a part of the energy of C_1 needs to be transferred. For this, it is enough to switch on $SW1$ for a time smaller than $T_1/4$.

Note that the actual voltage on the capacitor C_2 is negative with regard to the common node of the capacitors.

If C_1 is replaced with a voltage source, the circuit designer must fix the time τ for which $SW1$ is ON, so as to define the amount of energy transferred from the voltage source to C_2 in one cycle (Figure 9.4(a)). In this way, the inductor is connected to the voltage source for a time τ, and after τ, its current is equal to $i_{L1} = V_0\tau/L$, the transferred energy is equal to $W = i_{L\,max}^2 L/2$ and the actual voltage on C_2 is given by:

$$V_1 = \sqrt{\frac{L}{C_2}} i_{L\,max} = \frac{V_0\tau}{\sqrt{LC_2}} \qquad [9.4]$$

The switch $SW2$ should be ON exactly for the time necessary to discharge L on C_2, i.e. for the time:

$$\tau_1 = \frac{1}{4}T_2 = \frac{\pi}{2}\sqrt{LC_2}. \qquad [9.5]$$

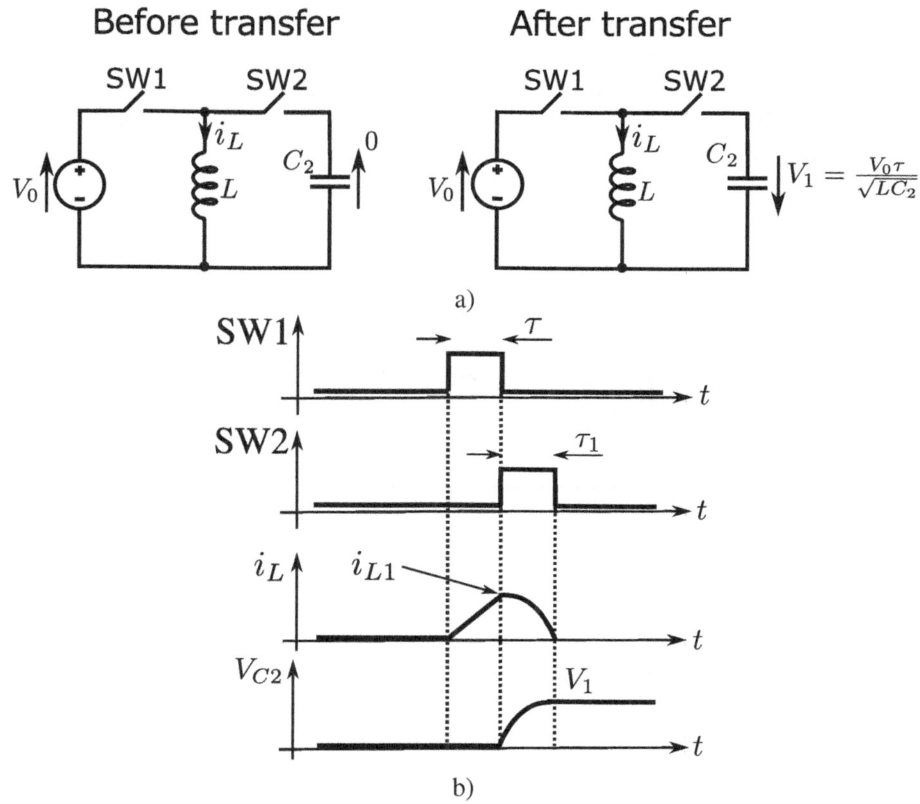

Figure 9.4. *Use of a DC-DC converter for lossless energy transfer between a voltage source and a capacitor. a) Schematic of the networks indicating the electrical state before and after the transfer and b) the timing diagrams illustrating the circuit operation*

Since this configuration is very common, we present detailed time diagram illustrating the operation of the circuit (Figure 9.4(b)).

The switching of $SW1$ and $SW2$ must be perfectly synchronized: $SW2$ must come ON exactly at the moment when $SW2$ comes OFF. If the switches are driven by an external command, this synchronization may be difficult to achieve. For this reason, often the switch $SW2$ is implemented with a diode, called sometimes flyback, flywheel or freewheel diode (Figure 9.5). At equilibrium, when SW1 is OFF and $i_L = 0$, the diode is negatively biased by the voltage V_{C2}, and it is cut off. When SW1 is ON, the diode is still negatively biased by the voltage $V_{C2} + V_0$. However, when $SW1$ is OFF and the current i_L in the inductor is positive (generated during the stage at which SW1 is ON), the inductor induces a voltage which biases the diode so it comes ON and it lets the inductor discharge on C_2. When the current of the inductor goes to zero and "tries" to become negative (this is a normal process of oscillation in a network composed of a capacitor and an inductor), the diode prevents it, and the circuit is locked in the state when C_2 stores the energy previously transferred from C_1 to L. The time diagrams and the evolution of the electrical quantities are exactly as in Figure 9.4, if the diode threshold voltage is neglected.

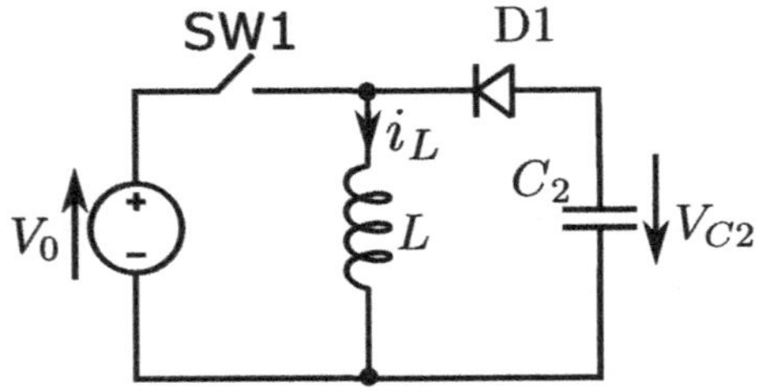

Figure 9.5. *Inductive DC-DC converter in which the switch SW2 is implemented with a flyback diode*

As we mentioned earlier, the voltages on the capacitors C_1 and C_2 measured with regard to the common node have opposite signs. For some applications, this may be inconvenient, and in this case a conventional BUCK DC-DC converter is used. The corresponding schematic is given in Figure 9.6. As $SW1$ is ON at $t = 0$ for some time $t = \tau$ as defined by the user, the voltage source charges both C_2 and the inductor through the network V_0LC_2. At $t = \tau$, $SW1$ is OFF and the inductor's induction voltage generates a positive biasing for the diode, so that the inductor's current continues to flow. The inductor discharges into C_2, and when the current becomes zero and starts to become negative, the diode cuts the circuit and remains negatively biased.

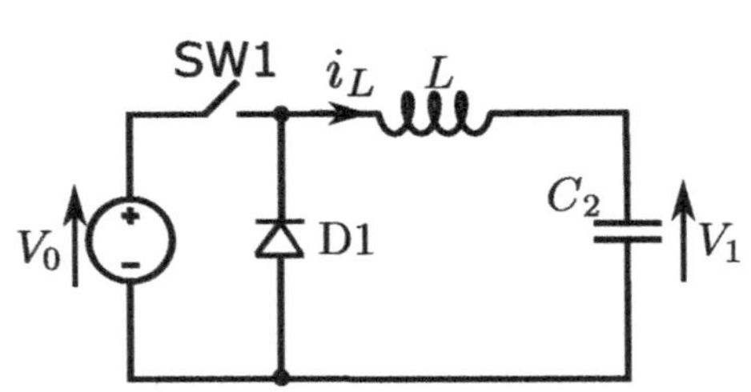

Figure 9.6. *Schematic of a BUCK DC-DC converter*

Contrarily to the circuit in Figure 9.2, the Buck converter is only able to generate a C_2 voltage *inferior* to the voltage V_0. This architecture is often used where it is required to generate a DC voltage lower than the voltage of an available source. In the capacitive harvesters, this topology may be used for the interface between the high-voltage conditioning circuit and a low-voltage load [DOM 13], as discussed in section 11.4.

We suggest the following exercise (which we leave for the readers): assuming V_0, C_2, L, τ and initial voltage $V_{2,0}$ of C_2 at $t = 0$ in Figure 9.5, find the energy transferred from the source V_0 to C_2, the peak current in the inductor, and the maximum value of τ allowed in this case and the timing parameter τ_1. An interesting variation of this exercise is when a capacitor C_1 is precharged to V_0 is used instead of the voltage source.

9.1.4. *Use of a variable capacitor*

Here, we consider a transfer of energy between a voltage source V_0 and a variable capacitor *initially charged to the same voltage*. For the transfer, the voltage source and the variable capacitor are connected (Figure 9.7(a)). If the capacitance changes slowly over time, the energy of the capacitance changes as:

$$W = \frac{C(t)V_0^2}{2} = \frac{Q(t)V_0}{2}, \qquad [9.6]$$

where $Q(t)$ is the instantaneous charge on the variable capacitor. The rate (power) at which this energy changes is equal to:

$$\frac{\mathrm{d}W}{\mathrm{d}t} = \frac{V_0}{2}\frac{\mathrm{d}Q}{\mathrm{d}t} = \frac{V_0 i(t)}{2}, \qquad [9.7]$$

where $i(t)$ is the current flowing through the capacitor and the voltage source. Since the electrical power provided by the voltage source V_0 is equal to $V_0 i(t)$, we can conclude that the voltage source gives or receives (depending on the sign of $C(t)$ variation) twice the energy received or given by the transducer. However, in the considered case, there is no dissipation: the energy difference represents the energy transferred between the mechanical and the electrical domain by the variable capacitor.

The voltage source can be replaced by a fixed capacitor C initially charged to the voltage V_0 (Figure 9.7(b)). In this case, the voltage on the two connected capacitors evolves according to the law:

$$V_C(t) = \frac{Q_0}{C(t)+C_1} = V_0\frac{C(0)+C_1}{C(t)+C_1} \qquad [9.8]$$

where $C(0)$ is the value of the variable capacitor at the beginning of the considered time interval (at $t = 0$), Q_0 is the initial total charge of the system equal to $V_0[C(0)+C_1]$. If $C_1 \gg C(t)$, the configuration is equivalent to the case where C_1 is implemented as a voltage source V_0.

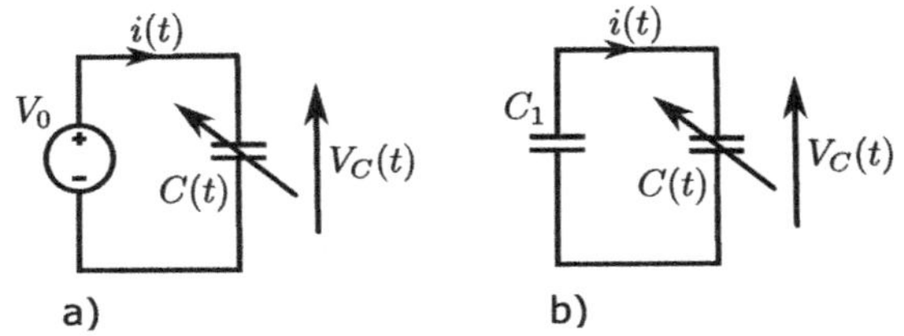

Figure 9.7. *Energy transfer between capacitors and voltage sources through the use of a variable capacitor*

In such a way, the considered network achieves a lossless exchange of energy between a variable capacitor $C(t)$ and a fixed capacitor C_1 or a voltage source. The exchange is not limited to the electrical domain: a part of energy is transferred to or from the mechanical domain. In the capacitive energy harvesters based on charge pumps considered in Chapters 10 and 11, this technique is used for the transfer of energy between the transducer and the energy reservoir, which may be a voltage source (a battery) or a large fixed capacitor.

In the examples we presented in sections 9.1.1 and 9.1.2, at the moment when two capacitors or a voltage source and a capacitor were connected in parallel, *their voltages were different*. We saw that such a configuration inevitably leads to the loss of a substantial part of energy, even if the parasitics are minimized. However, the use of a variable capacitor allows a regulation of the voltage at which two capacitors are connected, and as a result, such an energy transfer can theoretically be lossless. We emphasize this very important conclusion of this section: *a lossless transfer of energy between two capacitors without using an inductor requires equality of voltages on capacitors at the moment when they are connected in parallel by a switch.*

9.2. Conditioning circuits implementing triangular QV cycles

Two triangular QV cycles are known for conditioning of capacitive transducers in energy converters: the constant-voltage QV cycle and the constant-charge QV cycle. We now study conditioning circuits implementing these cycles.

9.2.1. *Constant-voltage conditioning circuit*

The constant-voltage QV cycle was discussed in Chapter 2. We showed that such a QV cycle provides the best energy yield achievable with a transducer having a given geometry, mechanical dynamics and supporting a given maximum voltage.

In this section, we discuss a practical implementation of a circuit achieving a constant-voltage QV cycle on a capacitive transducer. We highlight the fundamental practical difficulties, and explain the reasons of very limited use of this technique.

The scenario corresponding to the constant-voltage QV cycle and a corresponding electrical network is shown in Figure 9.8. The voltage source V_{max} can be a large fixed precharged capacitance. At the beginning of the cycle, C_t is discharged, SW is OFF:

1) When C_t is at its maximum, the transducer is precharged to a voltage V_{max} and immediately after that the switch SW goes ON connecting the transducer to the voltage source V_{max}. Note that the precharge of the transducer to V_{max} cannot be achieved by a simple switching ON of SW: a lossless mechanism of energy transfer should be employed (e.g. a DC-DC conversion, see section 9.1.3), and only when transducer voltage is V_{max}, the switch can become ON, so fixing the voltage on the transducer for all the time period during which C_t decreases. The DC-DC conversion used in this step is denoted as "conversion 1" in Figure 9.8.

2) When C_t decreases from C_{max} to C_{min}, the transducer is connected to the voltage source (SW is ON). During this time, the voltage source receives charges and energy from C_t, according to the principle described in section 9.1.4.

3) When C_t is at its minimum, the switch SW becomes OFF, and the transducer must be discharged, i.e. its energy should be transferred either to the load, or to the source V_{max}. In our example, the residual energy is transferred to the load; that is achieved by an inductive DC-DC conversion denoted as "conversion 2".

4) Since at each cycle the source V_{max} receives more energy than it delivers, the average energy of V_{res} increases with time. If the voltage source is implemented as a large capacitor, its voltage increases with time: that is not desirable, since, by initial hypothesis, V_{max} is the maximum voltage allowed in the given technology. The circuit should contain a mechanism removing energy from V_{max}, for example, by supplying a load. Since the load is supplied by a low voltage, an inductive DC-DC converter should be implemented, which is denoted as "conversion 3".

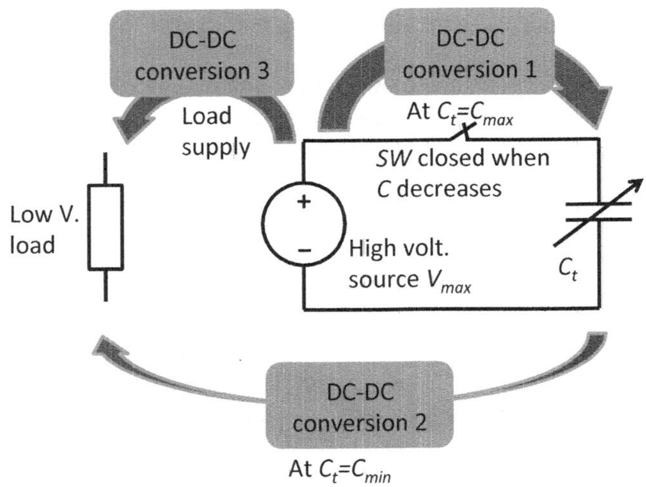

Figure 9.8. *Diagram illustrating the principle of operation of a conditioning circuit implementing a constant-voltage QV cycle*

This operation must be precisely synchronized with the variation of C_t. In this description, we do not consider how to implement such a synchronization.

Summarizing, an implementation of a constant voltage scenario requires:

– a high-voltage source;

– three inductive DC-DC conversions at each cycle;

– a switch connecting C_t and V_{max};

– a precise synchronization of the DC-DC conversions with the variation of C_t.

This implementation is very challenging because of the small available energy for the operation of the control electronics.

A successful implementation of a constant-voltage conditioning circuit was done by a research group of Georgia Tech University [TOR 09]. In this work, a rechargeable battery was used as V_{max} voltage source; the nominal battery voltage was up to 4.2 V. The implemented circuit generates a net power delivered to the battery of 86 nW, with a macroscopic transducer whose capacitance varied between 165.8 and 967.7 pF with frequency 30 Hz.

9.2.2. *Constant-charge conditioning circuits*

The constant-charge energy conversion scenario is similar with the constant-voltage scenario, except when the transducer capacitance decreases, the circuit fixes the charge of the transducer, and not the voltage. The corresponding aspect of QV cycle is given in Figure 8.1, (cycle OCB). The energy converted by the cycle is given by the formula:

$$\Delta W = \frac{1}{2}V_0^2 C_{max}\left(\frac{C_{max}}{C_{min}} - 1\right) = \frac{1}{2}V_{max}^2 \frac{C_{min}}{C_{max}}(C_{max} - C_{min}). \quad [9.9]$$

where V_0 is the voltage on C_t when $C_t = C_{max}$.

If we compare this formula with equation [2.33], it is evident that at the same V_{max}, C_{max} and C_{min}, the constant-charge QV cycle converts C_{max}/C_{min} times less energy than the constant-voltage QV cycle. This can also be seen from the plot of Figure 8.1.

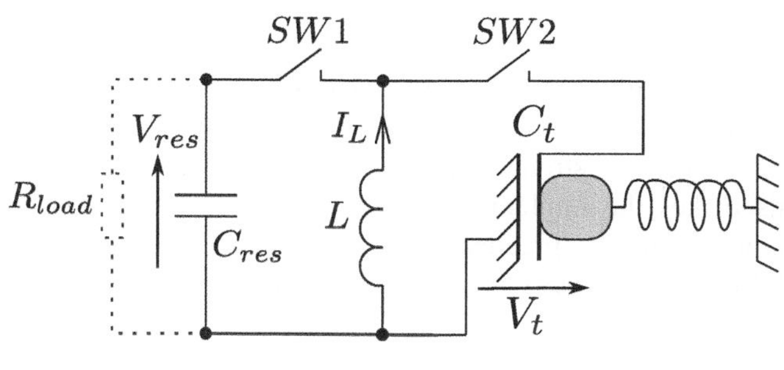

Figure 9.9. *Implementation of a constant-charge QV cycle*

However, a conditioning circuit implementing a constant-charge QV cycle is simpler than for a constant-voltage QV cycle. A common schematic of such a circuit is given in Figure 9.9. The schematic is composed of a large capacitor C_{res} initially precharged to the voltage V_{res}, or of a voltage source generating the same voltage, of an inductor and of two externally controlled switches. This topology was discussed in section 9.1.3. The operation scenario is the following:

– when C_t is at its maximum and discharged, a DC-DC converter transfers charges from a reservoir capacitor C_{res} to C_t *without loss of energy*, so that C_t has a charge Q_0 and a corresponding voltage $V_0 = Q_0 C_{max}$. The initial voltage V_0 is determined by the timing of the DC-DC conversion, and not by the voltage V_{res}. An appropriate command of the switches SW1 and SW2 is required;

– both switches are OFF, the transducer capacitance reduces while the charge of the transducer remains constant. The internal electrical energy of C_t increases;

– when C_t is at its minimum, the DC-DC converter transfers the charge back from C_t to C_{res}, without energy losses. The capacitor C_{res} receives a greater energy that it initially has given to C_t;

– the transducer capacitance increases while its charge is zero (the switches are OFF), until it reaches the maximum, and the cycle repeats.

In this way, the conditioning circuit operates as a bidirectional DC-DC converter. Two advantages over the constant voltage circuit should be noted:

1) The voltage V_{res} may be a low voltage, easy to generate. Nevertheless, the DC-DC converter is able to generate a higher initial voltage V_0 on C_t, if it is required for optimal energy conversion by the transducer (see discussion on inductive energy transfer in section 9.1.3).

2) The voltage V_{res} being low, it is compatible with the requirement of the load supply. In this way, one DC-DC conversion is removed (the "conversion 3" in Figure 9.8).

As was the case with the circuit implementing a constant-voltage QV cycle, a need for a precise synchronization of the DC-DC conversion with the variation of the transducer capacitance is a strong negative point of the constant-charge conditioning scenario. This makes the implementation of the control electronics complex and energy-consuming. Study of practical implementation of this solution can be found in [MEN 01], [MUR 04], [DES 05a], [DES 05b] and [BOI 13].

9.2.3. *Analysis of the circuit implementing a constant-charge QV cycle*

In this section, we consider in detail how the switches SW1 and SW2 in circuit of Figure 9.9 should be controlled so as to implement the constant-charge QV cycle. The explanations are illustrated in the diagram given in Figure 9.10.

We start the analysis from the state when $C_t = C_{max}$, $V_t = 0$, $V_{res} = V_{res}$, $i_L = 0$. We consider C_{res} large enough so as to neglect the variation on the voltage V_{res} at the time scale of one cycle.

When C_t is at its maximum, the transducer should be charged to some voltage V_0. The DC-DC converter operates as follows.

The switch $SW1$ is ON for some time τ, defined by the control circuits. If $V_{res} = const$, it is easy to show that the current in the inductor at $t = \tau$ is given by:

$$i_{L1} = \frac{V_{res}\tau}{L} \qquad [9.10]$$

and the energy W_0 transferred to the inductor is equal to:

$$W_0 = \frac{i_{L1}^2 L}{2} = \frac{V_{res}^2 \tau^2}{2L}. \qquad [9.11]$$

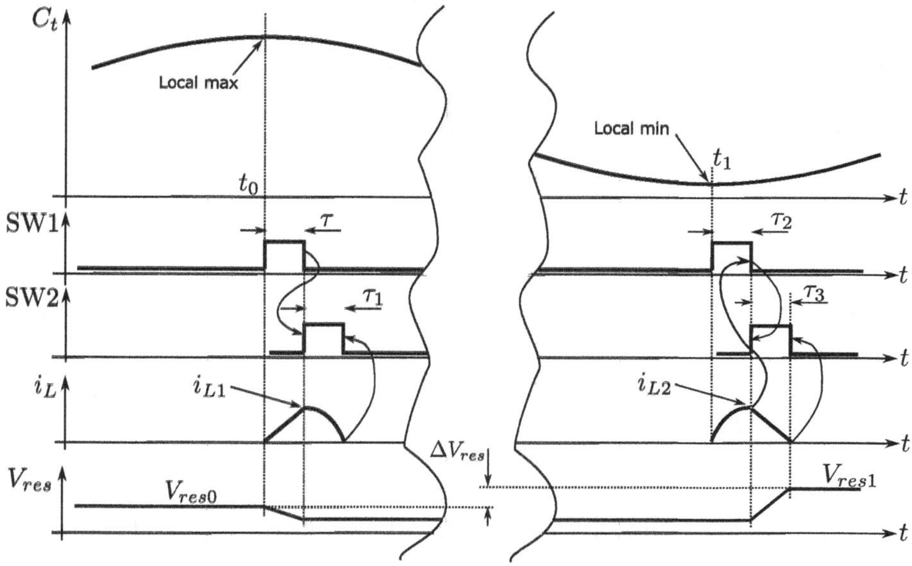

Figure 9.10. *Timing diagram explaining the operation of the constant-charge conditioning circuit in Figure 9.9*

The transfer of energy is lossless, so that W_0 is also the energy spent by C_{res}. At $t = \tau$, SW1 comes OFF, and SW2 comes ON, so that the inductor discharges on C_t. When i_L reduces to zero, the transducer has now received the energy W_0. The duration of this transfer is a quarter of the period of oscillation of the network LC_{max}, i.e. $\tau_1 = 0.25 \cdot 2\pi\sqrt{LC_{max}}$. Note that for the calculation of τ_1, we considered that $C_t = C_{max}$, i.e. we neglected the variation of C_t in time. This approximation is only valid if τ_1 is much less than the period of C_t variation. According to the principle of operation of a constant-charge circuit described in the previous section, the transfer of energy between C_{res} and C_t should be very fast, comparing to the time scale of C_t variation. After the transfer, the transducer voltage is:

$$V_0 = \sqrt{\frac{2W_0}{C_{max}}} \qquad [9.12]$$

and its charge is:

$$Q_0 = C_{max} V_0 \qquad [9.13]$$

At $t = \tau + \tau_1$, the switch $SW2$ must turn OFF, and the transducer is disconnected from the circuit during the decrease of C_t from C_{max} to C_{min}. A reverse process of transfer of energy from C_t to C_{res} starts at the moment t_1, when $C_t = C_{min}$. The switch $SW2$ comes ON for the time necessary for C_t to give all its energy to L: this is the time interval $\tau_2 = 0.25 \cdot 2\pi\sqrt{LC_{min}}$. This energy is now:

$$W_1 = \frac{Q_0^2}{2C_{min}} = W_0 \frac{C_{max}}{C_{min}} \qquad [9.14]$$

At time $t = t_1 + \tau_2$, the transducer is discharged, and the current of the inductor is given by:

$$i_{L2} = \sqrt{\frac{2W_1}{L}} \qquad [9.15]$$

At time $t = t_1 + \tau_2$, the switch SW2 must go OFF and the switch SW1 must come ON, so as to transfer the energy of inductor into C_{res}. This transfer takes a time τ_3, equal to:

$$\tau_3 = \frac{Li_{L2}}{V_{res}}. \qquad [9.16]$$

At the end of the cycle, the energy of C_{res} increases by:

$$\Delta W = W_1 - W_0 = W_0 \left(\frac{C_{max}}{C_{min}} - 1 \right) = \frac{V_{res}^2 \tau^2}{2L} \left(\frac{C_{max}}{C_{min}} - 1 \right), \qquad [9.17]$$

In this analysis, we neglected the variation of the voltage V_{res} at the time scale of one cycle. In reality, since the energy of C_{res} increases, there is an increase in its voltage, which we propose to calculate. If at $t = t_0$ the voltage of C_{res} is V_{res0}, at $t = t_1 + \tau_2 + \tau_3$ the voltage is V_{res1}. In order to calculate V_{res1}, we should consider the evolution of the energy of C_{res} between the times $t = t_0$ and $t = t_1 + \tau_2 + \tau_3$:

$$\frac{C_{res}V_{res1}^2}{2} = \frac{C_{res}V_{res0}^2}{2} + \Delta W. \qquad [9.18]$$

and we obtain:

$$V_{res1} = \sqrt{V_{res0}^2 + \frac{2\Delta W}{C_{res}}}. \qquad [9.19]$$

If C_{res} is large, the last formula can be approximated as:

$$V_{res1} \approx V_{res0} + \frac{\Delta W}{C_{res}}, \qquad [9.20]$$

so that the increase in the voltage of C_{res} after one cycle is:

$$\Delta V_{res} = V_{res1} - V_{res0} = \frac{\Delta W}{C_{res}} = \frac{V_{res}^2\tau^2}{2LC_{res}}\left(\frac{C_{max}}{C_{min}} - 1\right). \qquad [9.21]$$

9.2.4. *Practical implementation*

The switches SW1 and SW2 are activated twice by cycle, so that four control pulses should be generated "at right moments". In this section, we indicate some hints about how practically switches SW1 and SW2 can be controlled and implemented. Figure 9.11 illustrates the explanations:

DC-DC conversion 1: ($C_{res} \rightarrow C_t$ when $C_t = C_{max}$).

When a local maximum on C_t is detected, the switch SW1 is ON for duration τ. Implementation of this command requires a detector of local maximum of the transducer capacitance and a simple monostable circuit generating a pulse with fixed duration. Synchronously with the end of this pulse, the switch SW2 should turn OFF: another pulse should be generated, for the command of the switch SW2. This pulse should last till the moment at which the current in the inductor becomes 0. An easy way to implement the operation of SW2 is to replace it by a diode D3. Indeed, the diode is cut when SW1 is ON (its voltage on the diode is $-(V_t + V_{res})$, and when SW1 is OFF, the inductor L generates a voltage so as to push its current into the diode. When the inductor current reaches zero, the current "tries" to become negative, but the diode prevents this to happen, and the LC network is cut off. When the current in the inductor is zero, the diode is biased to $-V_t$, and it remains cut.

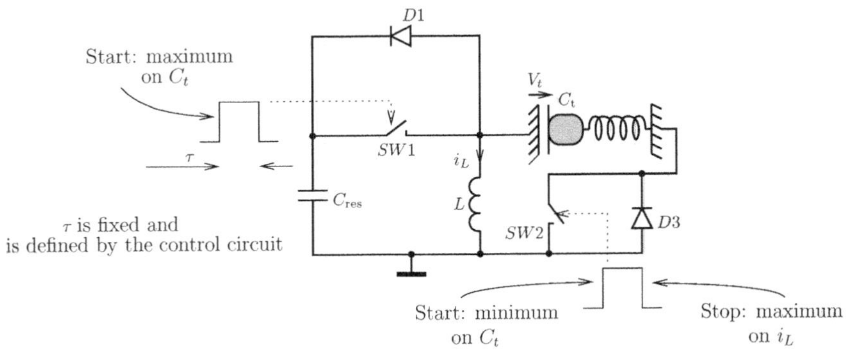

Figure 9.11. *A possible practical implementation of the constant-charge conditioning circuit whose principle is given in Figure 9.9*

DC-DC conversion 2: ($C_t \rightarrow C_{res}$ when $C_t = C_{min}$).

The implementation is very similar with the DC-DC conversion 1. When $C_t = C_{min}$, the switch SW2 must come ON (it is inserted in parallel with the diode). The switch should be cut when the voltage on C_t is zero, or when i_L is maximum. At the same time, the switch SW1 should come ON, and when the inductor is discharged (its current is zero), the switch SW1 should come OFF. The operation of the switch SW1 may be implemented with a diode D1, in a way similar with the DC-DC conversion 1. This diode should be insert in parallel with the switch SW1 used for the implementation of the DC-DC conversion 1.

In this way, we obtain the circuit presented in Figure 9.11, with timing diagram given in Figure 9.10. The following three functions should be implemented in order to make possible the described operation of the conditioning circuit:

– detection of C_{max};

– detection of C_{min};

– detection of a zero voltage on C_t.

These functions are not trivial to implement in practice, especially the detection of the local extrema of a variable capacitance. Moreover, a drawback of the architecture of Figure 9.11 is the use of a floating switch SW1[2]. It is possible to swap the positions

2 A dipole is called "floating" if none of its terminal is connected to the ground. Similarly, we can talk about a "floating voltage".

of C_{res} and the parallel connected network D1 and SW1, but in this case the reservoir capacitor becomes floating. Both the options are practically inconvenient. A possible solution to avoid is to replace the inductor in Figure 9.9 with a power transformer, as in Figure 9.12: in this case, both the switches can be referenced to the ground. The transformer plays the same role as the inductor: it is an intermediate reservoir for the energy during the transfer between C_{res} and C_t, but it offers a galvanic isolation between the networks of the transducer and the reservoir capacitor. In addition to having both switches referenced to the ground, this configuration offers the advantage of isolation of the low-voltage network connected to the load from the high-voltage network connected to the transducer. The timing and operation of the network is very similar with those of the circuit given in Figure 9.9. General information about DC-DC converters using transformers can be found in any manual on power electronics, e.g. [ERI 07]. Very detailed study and analysis of the conditioning circuit employing a transformer can be found in the PhD report of G. Despesse [DES 05a] (in French), and in [DES 05b] and [BOI 13].

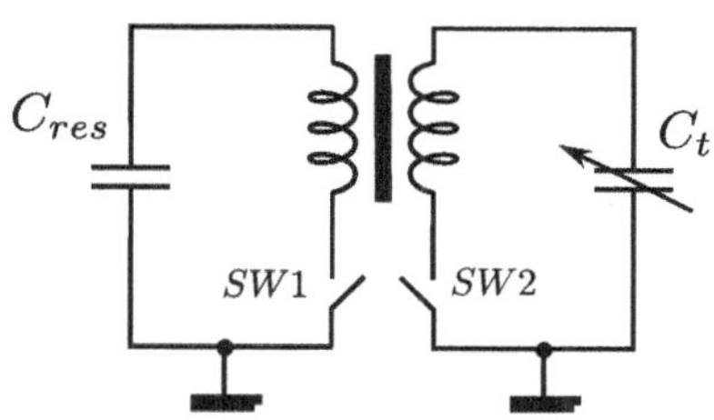

Figure 9.12. *Schematic of a conditioning circuit implementing a constant-charge QV cycle with the use of a transformer [DES 05b, BOI 13]*

9.3. Circuits implementing triangular QV cycles: conclusion

The conditioning circuits presented in this chapter are free from the limitations inherent to the continuous conditioning circuit discussed in Chapter 8. The energy converted from the mechanical domain is used to regenerate the charges on reservoir capacitor. Depending on the needs of the circuit operation, this allows a compensation of the energy consumed by the load or by the leakage, as well as a voltage increase in the reservoir capacitor.

The particularity of the conditioning circuits implementing triangular QV cycles is the need of an active external synchronization of the charge flow on the transducer with the motion of the transducer mobile electrode. We employ the term "active external synchronization" in order to emphasize the fact that the control of the switches $SW1$ and $SW2$ must be generated by supplementary blocks, and cannot be automatically generated by the circuit. Moreover, the precision of the timing of switching is paramount for the energy conversion efficiency, and as a result, a precise electronics is needed to achieve this synchronization. No doubt, this is the main weakness of these circuits, which motivates investigation of alternative architectures.

10

Circuits Implementing Rectangular QV Cycles, Part I

Chapters 10–11 are devoted to the class of circuits implementing rectangular QV cycle (DFBE in Figure 8.1). Such circuits are free from the main drawbacks of the two families of circuits considered in Chapters 8 and 9: (1) they allow an accumulation of the converted energy and thus a self-increasing of its internal energy from a possibly small initial biasing, (2) the synchronization of the charge flow on the transducer with the mobile mass vibration is done automatically through a use of passive diode networks. This topic is very extensive, and we made a choice to split its presentation into two parts.

In section 10.1, we introduce the general theory which is common for all circuits implementing rectangular QV cycles. In sections 10.2–10.4, we present the circuits based on the charge pump first proposed by Roundy *et al.* [ROU 02] and provide their detailed analysis. In section 10.5, we introduce the conditioning circuits based on the Bennet's electricity doubler [DE 11].

Chapter 11 discusses a few advanced points related to the practical use of conditioning circuits implementing rectangular QV diagrams: conditioning of transducers biased by an electret layer, electromechanical coupling, aspects related to practical use of rectangular QV cycle conditioning circuits.

10.1. Study of the rectangular QV cycle

Priorly to study the circuit topologies, we propose a discussion about the common properties of these circuits, which are due to the geometry of the implemented QV cycle.

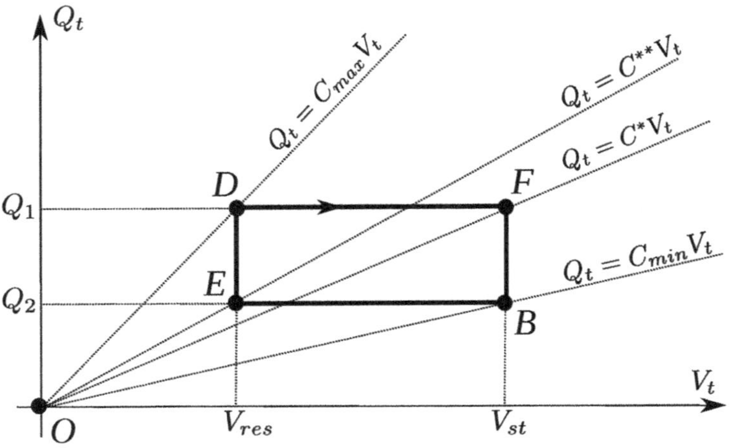

Figure 10.1. *Geometry of an ideal rectangular QV cycle*

Let us consider a rectangular QV cycle having the following parameters related to the cycle geometry (Figure 10.1):

– the extreme voltages of the cycle which we call V_{res}, V_{st}, such as $V_{res} < V_{st}$;

– the extreme charges of the cycle which we call Q_1, Q_2, such as $Q_1 > Q_2$;

– the four values of the transducer capacitance corresponding to four points of the rectangle forming the QV diagram[1]:

- $C_{max} = Q_1/V_{res}$ (corresponds to the point D),
- $C_{min} = Q_2/V_{st}$ (corresponds to the point B),
- $C^* = Q_1/V_{st}$ (corresponds to the point F),
- $C^{**} = Q_2/V_{res}$ (corresponds to the point E).

If a capacitive transducer is biased so that its electrical state follows the trajectory defined by such a cycle, the energy converted in one cycle is easily calculated as the area of the cycle:

$$\Delta W = (V_{st} - V_{res})(Q_1 - Q_2) = \\ = V_{res}^2 C_{max}\left(\frac{V_{st}}{V_{res}} - 1\right)\left(1 - \frac{V_{st}/V_{res}}{C_{max}/C_{min}}\right) \quad [10.1]$$

1 Each point (V, Q) on a QV diagram defines a unique value of the variable capacitor given by Q/V, otherwise, by the slope of the line connecting the point (V, Q) and the origin.

For this energy to be positive, it is required that:

$$\frac{C_{max}}{C_{min}} \geq \frac{V_{st}}{V_{res}}. \qquad [10.2]$$

It can easily be seen that the expression ΔW is a quadratic function of the variables V_{res}, V_{st}. As a result, if one of these voltages is fixed, there is an optimal value of other voltage. The readers can verify the following formulae:

– if V_{res} is fixed, the value of V_{st} maximizing ΔW is given by:

$$\frac{V_{st\ opt}}{V_{res}} = \frac{1}{2}\left(\frac{C_{max}}{C_{min}} + 1\right) \approx \left[\text{if } \frac{C_{max}}{C_{min}} \gg 1\right] \approx \frac{1}{2}\frac{C_{max}}{C_{min}} \qquad [10.3]$$

and the corresponding ΔW is:

$$\Delta W_{max}^{res} = \frac{1}{4}V_{res}^2 C_{min}\left(\frac{C_{max}}{C_{min}} - 1\right)^2 \approx \left[\text{if } \frac{C_{max}}{C_{min}} \gg 1\right] \approx \frac{1}{4}V_{res}^2 C_{max}\frac{C_{max}}{C_{min}} \qquad [10.4]$$

– if V_{st} is fixed, the value of V_{res} maximizing ΔW is given by:

$$\frac{V_{res\ opt}}{V_{st}} = \frac{1}{2}\left(\frac{C_{min}}{C_{max}} + 1\right) \approx \left[\text{if } \frac{C_{max}}{C_{min}} \gg 1\right] \approx \frac{1}{2} \qquad [10.5]$$

and the corresponding converted energy is

$$\Delta W_{max}^{st} = \frac{1}{4}V_{st}^2 \frac{C_{min}^2}{C_{max}}\left(\frac{C_{max}}{C_{min}} - 1\right)^2 \approx \left[\text{if } \frac{C_{max}}{C_{min}} \gg 1\right] \approx \frac{1}{4}V_{st}^2 C_{max} \qquad [10.6]$$

How practically can a rectangular QV cycle be implemented? It was mentioned in Chapter 1 that a conditioning circuit can be seen as a dipole (Figure 1.3). We now propose to identify the main properties of a dipole generating on the transducer a rectangular QV cycle, and to define its current–voltage characteristic.

By analyzing the QV diagram of Figure 10.1, it can be seen that when $V_{res} < V_t < V_{st}$, the transducer current is zero (since Q_t is constant). We conclude, that at this operation stage, the conditioning circuit presents a high impedance (an open circuit).

As the transducer voltage reaches V_{st} *when the transducer capacitance decreases* (the segment DF), the voltage of the transducer is fixed to V_{st} (the segment FB), i.e. the conditioning circuit presents a voltage source V_{st}. Note that the current of transducer is

$$i_t = \frac{dQ_t}{dt} = V_{st}\frac{dC_t}{dt} < 0 \qquad [10.7]$$

since $C_t(t)$ decreases. As a result, the transducer gives its charges (and hence its energy) to the conditioning circuit. The corresponding configuration is the same as studied in section 9.1.4.

A similar process happens when the transducer capacitance increases and its voltage reduces to V_{res} (the segment BE), while the capacitance continues to increase. In this case, the conditioning circuit behaves as a voltage source V_{res} (the segment ED). In this case

$$i_t = \frac{dQ_t}{dt} = V_{res}\frac{dC_t}{dt} > 0 \qquad [10.8]$$

since C_t increases, and as a result, the transducer takes charges and energy from the conditioning circuit.

We can define the current-voltage characteristic (relating i_t and V_t) of the conditioning circuit as follows:

– if $V_{res} < V_t < V_{st}$, then $i_t = 0$; (open circuit);

– if $i_t > 0$, then $V_t = V_{res}$; (voltage source V_{res});

– if $i_t < 0$, then $V_t = V_{st}$; (voltage source V_{st}).

Figure 10.2(a) presents the plot of the corresponding current–voltage characteristic. Such a characteristic corresponds to a voltage limiter implemented by the network given in Figure 10.2(b), often used for the electrostatic discharge (ESD) protection of integrated circuits. The use of this circuit for implementation of a rectangular QV cycle in capacitive energy harvesters was first proposed by Roundy [ROU 02]. In the literature addressing energy harvesting, this circuit is usually represented as in Figure 10.3(a), and is called a charge pump. Indeed, when this circuit is connected to a variable capacitor with a periodic variation of the capacitance, the low-voltage source V_{res} gives charges, whereas the high-voltage source V_{st} receives the charges. In this way, the circuit creates a charge flow from a low potential V_{res} to a high potential V_{st}. Such a transfer requires an energy: this energy comes from the mechanical domain through the variation of the transducer capacitance. After a cycle, the source V_{st} receives more energy than delivered by the source V_{res}: the energy difference is given by equation [10.1].

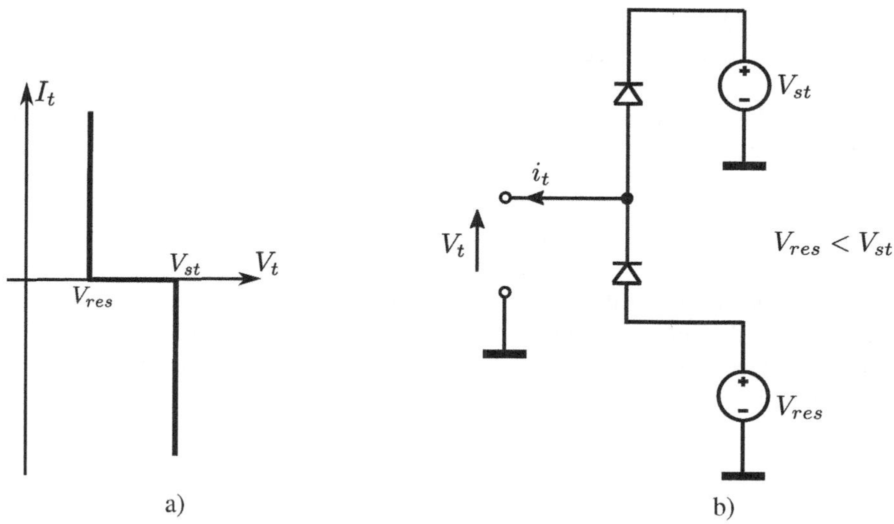

Figure 10.2. *Implementation of an ideal rectangular QV cycle: a) current–voltage characteristic of a dipole which, when connected to a variable capacitor, implements a rectangular QV cycle, b) an electrical network having such a characteristic*

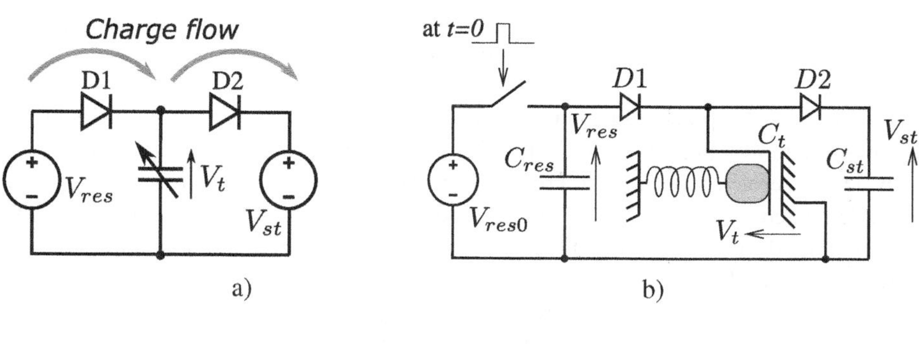

Figure 10.3. *Schematic of a charge pump implementing a rectangular QV cycle: a) ideal charge pump and b) practical implementation*

A great advantage of this circuit in comparison to a constant voltage or constant charge circuit is the automatic synchronization of the phases of the biasing with the variation of the transducer capacitance, without any need of external control. However, practical use of the charge pump in capacitive energy harvesters requires to answer the following questions:

– how to generate the voltages V_{res} and V_{st};

– how to transfer periodically the energy gained by the source V_{st} to some low-voltage storage device (a large capacitor or a battery).

In the following sections, we consider practical architectures of the charge pump which solve these problems.

10.2. Practical implementation of the charge pump

An implementation of the charge pump close to the described idealized circuit is obtained with the network of Figure 10.3(b). The voltage sources are implemented by charged fixed capacitances of large values. The values of C_{res} and C_{st} should be chosen so that:

$$C_{max} \ll C_{st} \ll C_{res}. \quad [10.9]$$

C_{res} capacitor is initially charged to a voltage $V_{res} = V_{res0}$, and so are C_{st} and C_t capacitors, because of the charge sharing through the diodes D1 and D2.

The capacitance C_{res} is large comparing to the two other capacitances of the network. It means when there is a charge flow between C_{res}, C_t and C_{st}, the voltage variation on C_{res} is weak in comparison with the voltage variation on the capacitors C_t and C_{st}. In the calculations of V_{st} below, the variation of V_{res} can be neglected and V_{res} is considered as a constant parameter.

By initially precharging the circuit to a voltage V_{res0}, we implement the configuration of Figure 10.3(a) with $V_{res} = V_{st}$, in which the converted energy per cycle is zero, according to [10.1]. However, since the V_{st} voltage source is implemented by a capacitor C_{st}, the charge transfer from C_{res} to C_{st} during the first cycle slightly increases the voltage across C_{st}. Hence, the corresponding QV cycle DBE is actually triangular, as shown in Figure 10.4. The next cycle starts in a configuration with a slightly larger V_{st}, yielding a trapezoidal QV cycle. After each cycle, V_{st} increases, so that all further cycles are different. It can be shown that the slope of the line DB, $F_1B_1, ...F_nB_n$ is $-C_{st}$ (steeper than C_{max}, but exaggerated in the figure): the shape of the cycles is close to rectangular, as $C_{st} \gg C_{max}$. The QV trajectory follows a succession of cycles, starting from $V_{st} \approx V_{res}$ and asymptotically approaching a degenerated cycle represented by a trapeze with a zero area (a horizontal line) corresponding to the V_{st} voltage given by:

$$V_{st\ sat} = V_{res} C_{max} / C_{min}. \quad [10.10]$$

This value is called the "saturation voltage" of the charge pump.

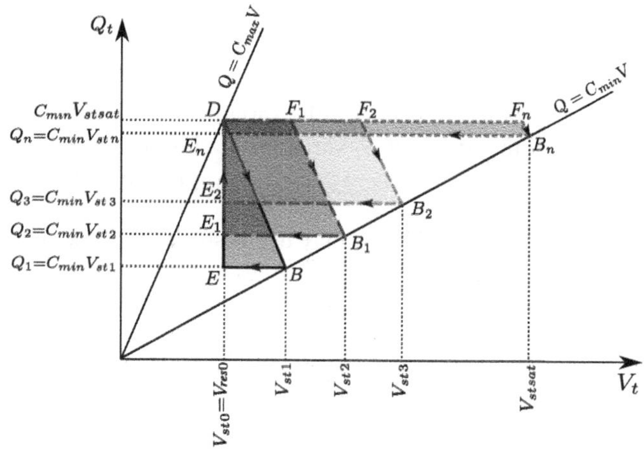

Figure 10.4. *Succession of charge-voltage cycles as the charge pump operates starting from the state when $V_{st} = V_{res}$: the first cycle DBE, the subsequent cycles $DF_1B_1E_1$, $DF_2B_2E_2$, ..., $DF_nB_nE_n$. If $n \to \infty$, the cycle is degenerated to a line passing through the points $D, F_1, F_2...F_n$. For a color version of figure, see www.iste.co.uk/basset/keh.zip*

As a result of the operation of this circuit, the capacitor C_{st} receives energy from C_{res}, and in addition, receives energy converted from the mechanical domain (the harvested energy). At every QV cycle of the charge pump operation, the overall electrical energy of the circuit increases. In what follows, we present a quantitative analysis of this process.

The voltage V_{st} at the end of the nth cycle evolves following the recursive equation [YEN 06]:

$$V_{st\ n} = \alpha V_{st\ n-1} + \beta, \qquad [10.11]$$

where

$$\alpha = \frac{C_{st}}{C_{st} + C_{min}}, \text{ and } \beta = V_{res}\frac{C_{max}}{C_{st} + C_{min}}. \qquad [10.12]$$

The solution of the discrete equation [10.11] is:

$$V_{st\ n} = \left(V_{st0} - V_{res}\frac{C_{max}}{C_{min}}\right)\left(\frac{C_{st}}{C_{st} + C_{min}}\right)^n + V_{res}\frac{C_{max}}{C_{min}}, \qquad [10.13]$$

where V_{st0} is the voltage on C_{st} capacitor at the zeroth cycle. In this example, it is assumed $V_{st0} = V_{res0}$.

Note that if C_t variations are periodic, the variable n is the operating time of the charge pump divided by the period of C_t variation, T_e. From [10.13], it can be derived that the asymptotic value of $V_{st\ n}$ as $n \to \infty$ is $V_{st\ sat}$ given by [10.10].

As V_{st} increases, the voltage V_{res} slowly decreases, as the amount of charges on C_{res} and C_{st} is constant (as we neglect the charges on C_t):

$$V_{res\ n} = V_{res0} - \frac{C_{st}}{C_{res}} \cdot (V_{st\ n} - V_{res0}) \approx V_{res0}\Big|_{\text{if } C_{res} \gg C_{st}}. \qquad [10.14]$$

An example of the evolution of V_{res} and V_{st} is given in Figure 10.5(a), where the charge pump starts from the state $V_{res} = V_{st} = V_t = V_{res0}$. From formulas [10.11] and [10.12], it can be seen that if $C_{st} \gg C_{max}$, the increment of V_{st} during every cycle is small, and the charge pump runs through cycles for all possible V_{st}, from V_{res0} to the saturation value $V_{st\ sat}$.

10.2.1. *Evolution of the harvested energy*

Neglecting the energy in the transducer capacitance since it is small, and supposing that the C_{st} voltage is initially V_{res0}, the energy harvested during the n first C_t variation periods is stored in the capacitors C_{res} and C_{st}, and is given by:

$$W_n = \frac{1}{2}C_{st}V_{st\ n}^2 + \frac{1}{2}C_{res}V_{res\ n}^2 - \frac{1}{2}(C_{res} + C_{st})V_{res0}^2, \qquad [10.15]$$

where $V_{res\ n}$ is the C_{res} voltage after n vibration cycles and V_{res0} is the initial voltage of C_{res}. At the same time, the sum of charges of C_{st} and C_{res} is constant (Q_0), so, the following relation holds (neglecting the charge shared with C_t):

$$Q_0 = (C_{st} + C_{res})V_{res0} = C_{st}V_{st\ n} + C_{res}V_{res\ n}. \qquad [10.16]$$

Using [10.16], equation [10.15] can be simplified:

$$W_n = \frac{C_{st}C_{res}}{C_{st} + C_{res}}(V_{st\ n} - V_{res\ n})^2/2. \qquad [10.17]$$

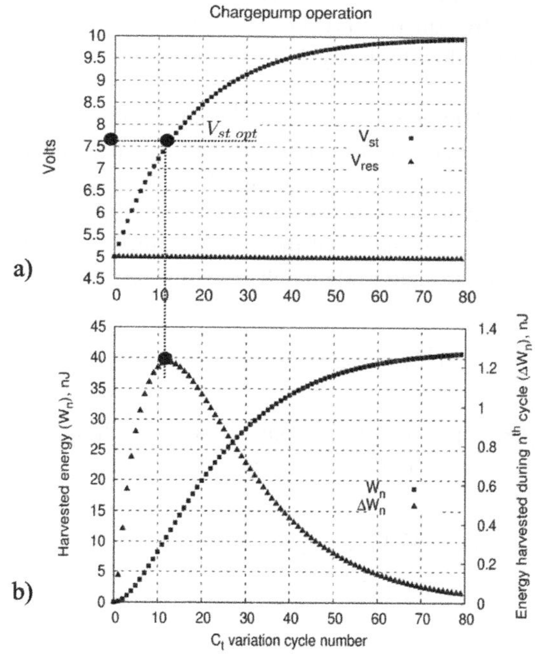

Figure 10.5. *Evolution of the electrical state of the charge pump without flyback. a) Evolution of V_{st} and V_{res} voltages, b) evolution of the total converted energy (W_n) and the energy converted at one cycle (ΔW_n). The simulation was done with $C_{min} = 200\,pF$, $C_{max} = 400\,pF$, $C_{st} = 3.3\,nF$, $C_{res} = 10\,\mu F$, frequency of C_t variation is 300 Hz*

This equation suggests that the harvested energy is stored in the capacitor composed of C_{st} and C_{res} connected in series. This formula can be further simplified if $C_{st} \ll C_{res}$ and if C_{res} is large, so that $V_{res\,n} \approx V_{res0}$. In this case, we can write:

$$W_n \approx C_{st}(V_{st\,n} - V_{res0})^2/2. \qquad [10.18]$$

This equation suggests that the converted energy is mainly stored in the C_{st} capacitor.

The value

$$\Delta W_n = W_n - W_{n-1}, \qquad [10.19]$$

gives the energy converted during one cycle. For the case when $C_{st} \gg C_{max}$, ΔW_n is equal to ΔW given by equation [10.1], if in [10.1] V_{st} and V_{res} are replaced by V_{res0} and $V_{st\ n}$ corresponding to the cycle.

In the formula [10.18] expressing the energy converted from the state where $V_{st} = V_{res0}$, there is no explicit dependence on discrete time n (the cycle number). If $C_{st} \gg C_{max}$, the variation of the voltage V_{st} is negligible at the time scale of one cycle. For this reason, we can state that V_{st} evolves continuously, and consider it as "macroscopic state parameters" defining the energy of the charge pump. It allows us to remove the indices n in equation [10.18] and to define the total harvested energy W_n as:

$$W(V_{res0}, V_{st}) = C_{st}(V_{st} - V_{res0})^2/2. \qquad [10.20]$$

If, during the operation of the charge pump, the V_{st} voltage increases from some V_1 to some V_2, the corresponding converted energy is given by:

$$\begin{aligned} \Delta W(V_{res0}, V_1, V_2) &= W(V_{res0}, V_2) - W(V_{res0}, V_1) \\ &= C_{st}(V_2 - V_1)(V_2 + V_1 - V_{res0})/2. \end{aligned} \qquad [10.21]$$

Figure 10.5(b) presents the plots characterizing the evolution of the energy in the charge pump as a function of the cycle number. During each cycle n, the total energy (square dots) increases by the value given by [10.1] calculated for the V_{st} resulting in the previous cycle (triangular dots). Only a few cycles convert a significant amount of energy: those corresponding to V_{st} values close to the optimal $V_{st\ opt}$, (see the analysis of section 10.1 for fixed V_{res}). This value is situated in-between the extreme values of V_{st}, and a freely running charge pump finishes by entering into the saturation mode, in which $V_{st} \approx V_{st\ sat}$ and $\Delta W \approx 0$.

10.3. Shortcomings of the single charge pump and required improvements

Because of the saturation, the charge pump alone is not useful for capacitive KEHs. Practical conditioning circuits should be able to (1) ensure a sustainable energy conversion and (2) accumulate the converted energy. These two very important points are discussed in this section.

10.3.1. *Need for a flyback*

The plot ΔW_n in Figure 10.5(b) may be interpreted as the time evolution of the average power converted in one cycle. The goal of a harvester is to permanently maintain the converted power close to a maximum level. For this reason, a charge pump needs periodic *flyback*: a mechanism returning some charges from C_{st} to C_{res} so as to keep V_{st} close to the optimal value and to prevent the charge pump from saturation. Possible implementations of the flyback will be discussed in section 10.4.

The action of the charge pump can be seen as the generation of an average current flowing from a low potential V_{res} to a higher potential V_{st} (Figure 10.6). This current is roughly equal to

$$i_{ch\ pump} = \frac{Q_1 - Q_2}{T_e} = \frac{V_{res}C_{max} - V_{st}C_{min}}{T_e}. \quad [10.22]$$

where Q_1 and Q_2 are the parameters of the QV cycle (see Figure 10.1), T_e is the period of the transducer capacitance variation.

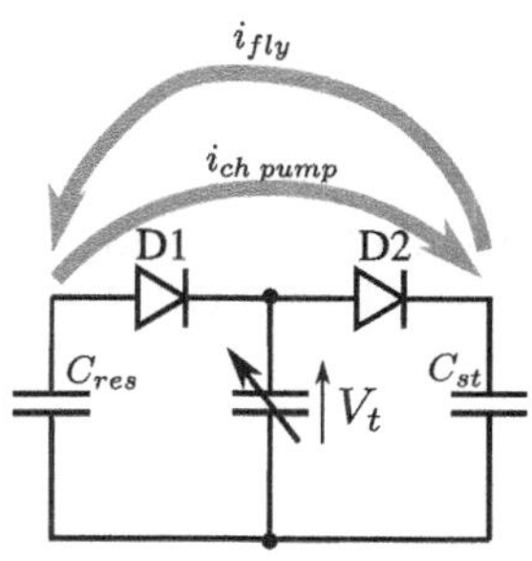

Figure 10.6. *Principle of operation of a charge pump with a mechanism of return of charges from C_{res} to C_{st}, called "flyback"*

The flyback must create an opposite current i_{fly} from C_{st} to C_{res}, preventing the accumulation of charges on C_{st}, and so as to avoid the charge pump saturation. This current, flowing from the high to low potential, *removes* the electrical energy from the charge pump, and in a practical application, this energy should be transferred to a useful load. In the steady-state mode, the rate at which the energy is removed from the charge pump is equal to the rate at which the energy is generated by the charge pump, i.e. to the power of the electromechanical energy conversion.

10.3.2. *Auto-increasing of the internal energy*

From [10.1] and further analysis of section 10.2, it can be seen that the energy converted by a charge pump is proportional to the square of the voltage V_{st} and V_{res}, i.e. proportional to the internal energy of the circuit in Figure 10.3(b). Practical studies highlight that dozens of volts may be needed for optimal operation [BAS 14].

However, the initial charge can only be obtained by a low-voltage battery existing in the system, or by a complementary piezoelectric KEH as proposed in [FLO 11]. In both cases, the initial voltage generated in this way will certainly be low (few volts).

As a result, the conditioning circuit should be able to use a part of the converted energy in order to increase its internal energy represented by the voltage on the largest capacitor V_{res}. We call this property "auto-increasing of the internal energy", and the corresponding operation mode "accumulative mode". This mode allows not only a high bias voltage for the circuit, but also a compensation of possible losses in the conditioning circuit (leakages in the capacitors).

The capability of auto-increasing the bias is not offered by the basic charge pump without flyback.

10.4. Architectures of the charge pump with flyback

10.4.1. *Resistive flyback*

The simplest way to implement a flyback is to connect a load between C_{res} and C_{st} (Figure 10.7), originally proposed in [FLO 11]. Here, the load is represented with a resistance R_{load} whose current is:

$$i_{fly} = \frac{V_{st} - V_{res}}{R_{load}}. \qquad [10.23]$$

A simplified analysis may be done by analyzing independently the charge flows of the charge pump and through the resistor, and then by superimposing them. The point at which both currents are equal to each other ($i_{ch\ pump} = i_{fly}$) corresponds to the steady-state operation. Although it is an approximation based on the averaging method [NAY 93], it gives good analytical results.

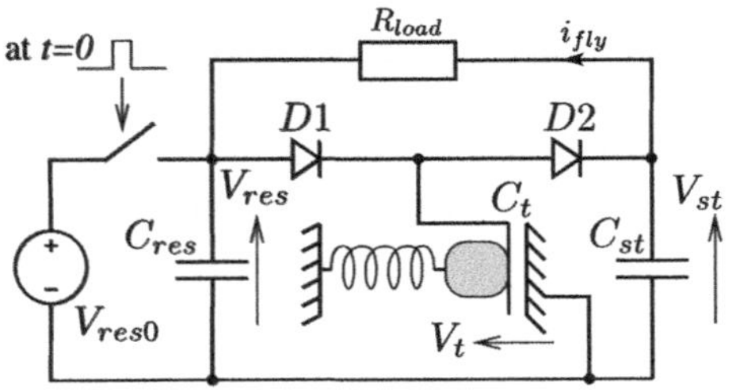

Figure 10.7. *Charge pump with resistive flyback*

In the steady state, the average V_{st} voltage on a period T_e is the same for all cycles. The stability of this mode can be proven by supposing a perturbation which yields, for example, $i_{ch\ pump} < i_{fly}$. In this case, the load current consumption overweighs the current due to the charge generation by the pump, and V_{st} decreases. This results in decreasing of i_{fly}, and thus in reducing the initial perturbation.

Under these considerations, the average steady-state voltage V_{st} with resistive flyback can be found from equations [10.22] and [10.23]. Conversely, the value of R_{load} yielding a desired V_{st} (for example, which yields a maximum ΔW) can be calculated.

The resistive flyback configuration provides a sustainable energy conversion by preventing the charge pump from reaching saturation. This is an efficient solution from the point of view of the simplicity of implementation. However, it does not allow an auto-increasing of its internal energy (see section 10.3.2), and for this reason, it is not usable in practical applications.

10.4.2. *Inductive flyback*

10.4.2.1. *Architecture and operation*

The inductive flyback is a Buck DC-DC converter, transferring energy from C_{st} to C_{res} (Figure 10.8(a)), whose general principle of operation was discussed in section 9.1.3. The inductive flyback operates in two steps:

1) When V_{st} is high so that the efficiency of the charge pump decreases, some external control device (see the explanation below) activates the switch SW, and the current in the inductor L starts to increase. This results in transfer of charges from C_{st} to C_{res} so that V_{st} decreases, and in transfer of the energy previously generated by the charge pump on C_{st} toward the inductor L.

2) As the value of V_{st} decreases, it becomes close to V_{res}. The same external control device cuts the switch, so that the inductor discharges through the flyback diode $D3$ onto C_{res}. This results in a transfer of the inductor energy to C_{res}, and an *increase* in the C_{res} voltage, since the diode $D3$ is connected to the ground. In this way, the converted energy is used to extract charges from the ground and inject new charges into the charge pump.

Capacitor C_{res} receives charges twice. In the first phase, it receives the charges previously given to C_{st} during the pump operation. In the second phase, it receives new charges from the ground. In this way, C_{res} receives the energy it has given to C_{st} during the operation of the charge pump, and receives the energy converted from the mechanical domain. If such an operation is cyclic, the average voltage and energy of C_{res} will grow. Hence, two problems are solved: (1) even if the capacitors are leaky, the losses are compensated by the charges extracted from the ground, (2) a large capacitor C_{res} contains the harvested energy available for the load (represented by dotted lines in Figure 10.8(a)). If no load is present, C_{res} accumulates the converted energy, and thus implements an "auto-increasing" or "accumulative" mode.

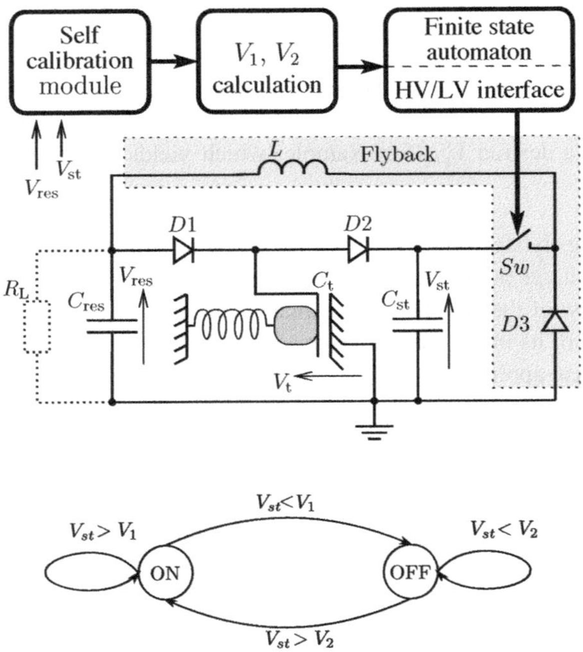

a)

b)

Figure 10.8. *Conditioning circuit with inductive flyback. a) The architecture of the circuit: the basic charge pump (as presented in section 10.2), the inductive flyback (gray background, section 10.4.2) and the architecture of the adaptive control (see section 10.4.2), b) the finite state automaton explaining the algorithm of the flyback switch control*

The advantage of the charge pump with inductive flyback is the possibility of a precise control of the QV cycle corresponding to the energy conversion. This is achieved by a modulation of the average V_{st} voltage so as to remain in the optimal mode for energy conversion.

A drawback of this circuit is the need of an external command for the switch. However, this command is synchronized not with the mobile mass vibrations, but with the evolution of the voltage V_{st}: first, a voltage can easily be measured, and second, the evolution speed is low compared to the period of the capacitance variation. Hence, the power cost of such a command is smaller than for circuits implementing constant-voltage or constant-charge QV cycles (see sections 9.2.1 and 9.2.2).

10.4.2.2. *Practical implementation*

In order to take maximum advantage of the charge pump with inductive flyback, a "smart" management of the flyback is required (see functional representation in Figure 10.8(a)). This was first highlighted in [GAL 07]. The most important issue is the selection of an interval $[V_1, V_2]$ for V_{st} values, so that $V_1 < V_{st\ opt} < V_2$, and V_1 and V_2 are close to V_{opt}. As far as V_1 and V_2 are chosen, the operation of the switch can be controlled by a finite-state machine (Figure 10.8(b)), and the evolution of V_{st} and the average power is given in Figure 10.9.

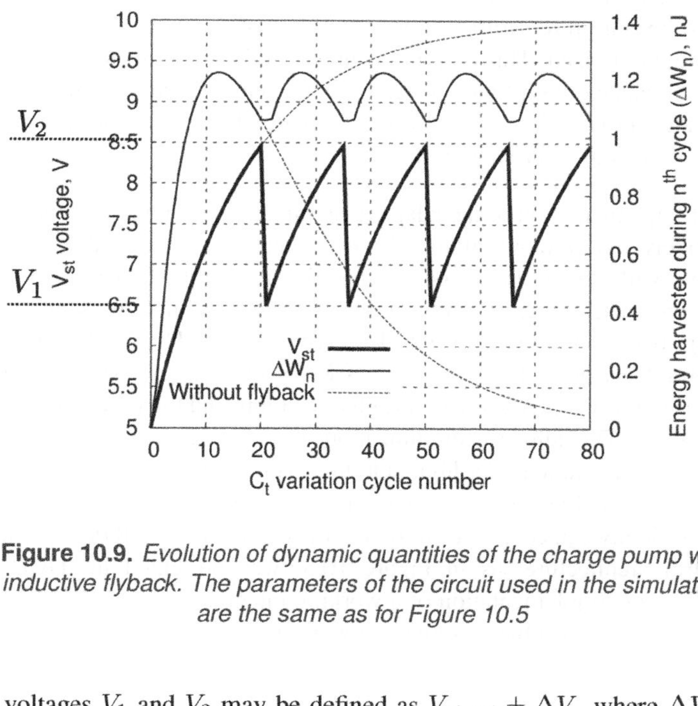

Figure 10.9. *Evolution of dynamic quantities of the charge pump with inductive flyback. The parameters of the circuit used in the simulation are the same as for Figure 10.5*

The voltages V_1 and V_2 may be defined as $V_{st\ opt} \pm \Delta V$, where ΔV is chosen large enough so as to minimize the frequency of the switch activation, but at the same time small enough to keep the average power close to the maximum value. By considering [10.13], the readers can verify that the value of the capacitance C_{st} controls the number of cycles during which the voltage V_{st} goes from V_1 to V_2, and at the same time, the energy that the flyback circuit must transfer from C_{st} to C_{res}. A trade-off must be found, as a function of the practical implementation of the whole system.

However, $V_{st\ opt}$ is not directly measurable, and this is a fundamental problem. From [10.3], it can be seen that $V_{st\ opt}$ depends on C_{max}/C_{min}, which is not known *a priori* in practice, as it depends on the amplitude of the mobile mass vibration (see discussion on the mechanical coupling in section 6.2). One of the possible ways to determine $V_{st\ opt}$ is to measure the saturation voltage of a freely running charge pump. This voltage is given by $V_{st\ sat} = V_{res}C_{max}/C_{min}$. As it can be seen in Figure 10.9, the values of V_1 and V_2 can be chosen in-between V_{res} and $V_{st\ sat}$, for example,

$$\begin{aligned} V_1 &= V_{res} + 1/3 \cdot (V_{st\ sat} - V_{res}), \\ V_2 &= V_{res} + 2/3 \cdot (V_{st\ sat} - V_{res}) \end{aligned} \qquad [10.24]$$

As far as $V_{st\ sat}$ is known, the calculation on V_1, V_2 may be done by simple analog circuits, such as a voltage divider [DUD 14]. To determine $V_{st\ sat}$, the charge pump should be run up to saturation periodically, and the final V_{st} value should be measured and memorized. This measurement should be periodic, since the vibration amplitude may change with time. We call this technique "auto-calibration", since a periodic recalculation of V_1 and V_2 allows an optimal operation of the circuit for a range of parameters of external vibrations.

Practical implementation of the adaptive architecture in Figure 10.8 is challenged by the relatively low power available for the control circuitry, and by the need to manage high voltages in the conditioning circuit. Most studies used an "old" complementary metal-oxide semiconductor (CMOS) technology supporting high voltages (e.g. 0.7 μm CMOS in [KEM 13]). The work of Dudka [DUD 12] used a mixed high-voltage/low-voltage 0.35 μm CMOS technology, where high-voltage circuits are used for the interface with the charge pump, and the processing is done by low-voltage parts. This allows a minimization of the power overhead of the control electronics. Implementation of a high-side high-voltage metal oxide semiconductor (MOS) switch for the flyback control with a low-voltage control interface represents a particular difficulty. An advanced study of a fully integrated CMOS implementation of the control architecture, as in Figure 10.8, was presented in [DUD 14] and [DUD 12]. To date, implementation of a working integrated circuit (IC) prototype of this architecture is still a subject of ongoing work in several research groups.

10.5. Conditioning circuits based on the Bennet's doubler

10.5.1. *Introduction of the principle*

Introduced at the end of the 18th Century [BEN 87], the Bennet's electricity doubler is one of the first devices allowing a measure of a voltage, by amplifying the induced charge. It has recently been proposed by de Queiroz [DE 11] for capacitive kinetic energy harvesters. Since then, several works have further developed this concept, adapting it to microscale KEH [DRA 13, RIS 14, DOR 14, LEF 15].

The Bennet's doubler is a switched-capacitor network, whose goal is a steady separation of the electrical charges and an accumulation of the separated charges in the capacitors. A steady accumulation of the charges may lead to very high (theoretically unlimited) voltages. Before analyzing a conditioning circuit based on the Bennet's doubler, we propose an analysis of a more simple network widely used as a serial-parallel switched capacitors DC-DC converters. The purpose of the proposed discussion is to help the readers to acquire an intuitive understanding of the operation of the Bennet's doubler-based conditioning circuit, whose detailed and rigorous analysis will be given in the next section.

Consider an AC current source providing at the first half period a charge ΔQ, and pulling at the second half period the same charge. If during the first half period such a source is connected to a series capacitive network composed of two identical capacitors initially charged to identical voltages (Figure 10.10(a), $V_1 = V_2$), each capacitor receives a charge ΔQ. However, if during the second half period, the network topology is reconfigured and the previously charged capacitors are connected in parallel (Figure 10.10(b)), the current source takes a charge $\Delta Q/2$ from each capacitor. At the end, each capacitor receives a charge $\Delta Q/2$, which is added to the pre-existing charges of the capacitors. The cyclic reconfiguration of the circuit topology from series to parallel is usually achieved by externally controlled switches. In this case, the circuit can be seen as an AC-DC converter, since the AC current with finite maximum and minimum values is converted to a DC voltage, whose value is controlled by the number of operation cycles.

In order to achieve an automatic cyclic reconfiguration of the circuit between the topologies of Figures 10.10(a) and (b) (i.e. without use of an external control), we may benefit from the fact that in each topology the current has a specific direction. Figure 10.10(c) presents a topology in which this reconfiguration is automatic due to a diode network. When the current is positive, the diode D2 is ON (series configuration), and when the current is negative, the diodes D1 and D3 are ON (parallel configuration). The use of the diodes introduces a new feature to the network: the voltages across C_1 and C_2 may be different (for example, because of the initial charging). In this case, when the current is negative, all the current passes through the branch with the maximum voltage, and the current source generates a voltage $max(V_1, V_2)$.

The requirement applied to the current source is to be able to generate a voltage[2] $V_1 + V_2$ during the positive current phase, and a voltage $\max(V_1, V_2)$ during the negative current phase. This is always possible for an *ideal* current source, which, by definition, can generate any voltage necessary to ensure the required current value.

2 Recall that an ideal current source generates in the external network connected to a current of a given intensity. For this, it generates a *voltage* necessary to fix such a current. The value of this voltage is determined by the external circuits. For instance, if a resistance R is connected to a current source generating a current I, the source generates a voltage RI.

This is not a case of a real current source (we will see it in section 10.5.2, when a capacitive transducer will be associated with this circuit).

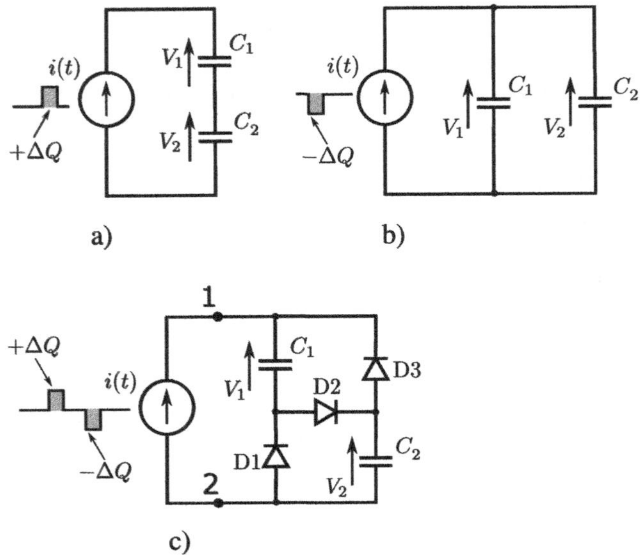

Figure 10.10. *Principle of the charge doubling in Bennet's doubler: a) series configuration of capacitors when the current is positive, b) parallel configuration of capacitors when the current is negative, c) a diode network allowing an automatic switching between the configurations, depending on the sign of the current*

Note that on the time scale of one period, the network in Figure 10.10(c) implements a voltage limiter with the characteristic as in Figure 10.2(a) if C_1 and C_2 are large, and if ΔQ is small comparing to the charges of these capacitors. Indeed, in this case, the variations of the charges and hence of the voltages on C_1 and C_2 are negligible at the time scale of one period, and the current source voltage is $\max(V_1, V_2)$ when the current is negative, and $V_1 + V_2$ when the current is positive. If, for some reason, the voltage applied to the terminals 1 and 2 of the network in Figure 10.10(c) is in-between these limits, all diodes are blocked and the current is zero.

A similarity of the considered network (Figure 10.10(c)) with a voltage limiter suggests that it can be used for implementation of a conditioning circuit achieving a rectangular QV cycle. Such a configuration is shown in Figure 10.11, and was originally proposed by de Queiroz [DE 11]. Its operation principle can be understood if we consider a functional similarity between a variable capacitor $C_t(t)$ and a current source. Consider a variable capacitor whose capacitance $C_t(t)$ reduces in time. When connected to a voltage source V, it generates a positive current

$i = V\mathrm{d}C_t/\mathrm{d}t$: the similarity with a positive current source is obvious. When such a variable capacitor is connected to an open circuit, it generates an increasing voltage $Q_t/C_t(t)$, where Q_t is the charge of the transducer (which is constant in time, since the current is zero). If the C_t decrease is bounded by some C_{min}, the voltage increase is bounded by Q_t/C_{min}, and if $C_{min} \to \infty$, the voltage tends to become infinite. A current source connected to an open circuit generates an infinite voltage if the source is ideal, or a voltage limited by IR_S if the source is real, i.e. having an internal resistance R_S. There is hence a similarity in the electrical behavior between a capacitor having a decreasing/increasing capacitance and a positive/negative current source correspondingly.

As a conclusion, a Bennett's doubler conditioning circuit is a series-parallel switched capacitors AC-DC converter, where the input AC current is generated by a variable capacitor. As the analysis presented in the next section will show, this circuit does not experience a saturation like the charge pump in Figure 10.3(b), and as a result, does not need any external control for steady generation of electricity.

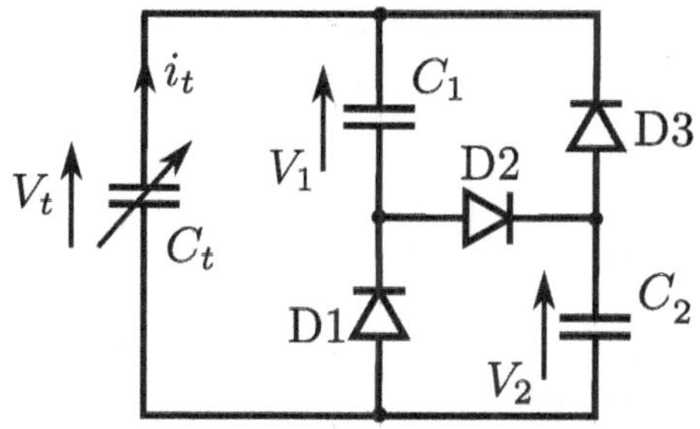

Figure 10.11. *Bennet's doubler used as a conditioning circuit for a capacitive KEH, as proposed in [DE 10]*

10.5.2. *Analysis of the Bennet's doubler conditioning circuit*

The goal of the analysis is a derivation of the evolution law of the electrical state of the Bennet's doubler-based conditioning circuits, considering the parameters of the circuit, the parameters of $C_t(t)$ variation and the initial conditions are known. This analysis provides the analytical expressions linking the rate of the energy conversion with the circuit parameters, and defines the restrictions applied on the circuit parameters in order to achieve the desirable mode of the energy conversion.

For the circuit analysis, we use the following conventions and hypotheses not restraining the generality of the study: (1) the capacitors $C_1 < C_2$, (2) when at the ith cycle C_t is at its maximum value C_{max}, the voltages V_1, V_2 and V_t have values named $V_{1\bar{i}}$, $V_{2\bar{i}}$ and $V_{t\bar{i}}$, the overline meaning that this value corresponds to the maximum value of C_t; similarly, when $C_t = C_{min}$, the voltages are named with underlined index i, (3) an ith cycle starts when $C_t = C_{max}$, so that the sequence of the values of, for example, V_t is: $(V_{t\bar{i}} \to V_{t\underline{i}} \to V_{t\overline{i+1}}...)$, (4) at the moment when $C_t = C_{max}$, the voltage V_t is equal to the maximum voltage among V_1 and V_2. The objective of the analysis is to find the evolution of the voltages on the three capacitors during one cycle, and to see how the energy of the capacitors evolves from one cycle to another.

Two cases should be considered: the steady-state operation, and the transient process. We will show that the electricity is generated both in the steady-state operation and during the transient process, and that the transient process always ends up by establishing the steady-state mode.

1) *Steady-state operation of the circuit*

We define the steady-state mode as follows: at each ith cycle, when $C_t = C_{max}$, the voltages on the three capacitors are equal to some voltage $V_{\bar{i}}$:

$$V_{1\bar{i}} = V_{2\bar{i}} = V_{t\bar{i}} = V_{\bar{i}}, \qquad [10.25]$$

and the sequence $V_{\bar{i}}$ is strictly increasing, i.e. $V_{\overline{i+1}} > V_{\bar{i}}$.

In order to describe the operation of the circuit in the steady state and to derive the necessary and sufficient conditions of the steady-state operation, we study the evolution of an ith cycle starting from the configuration given by [10.25]. The goal of the analysis is to check that if the equality [10.25] is true for some cycle i, then it is also true for the subsequent cycles.

C_t decreases from C_{max} to C_{min}. The voltage of the transducer starts to increase, and it becomes greater than $V_{\bar{i}}$. As the transducer voltage is under $2V_{\bar{i}}$, the diodes remain OFF, and the transducer keeps its charge equal to $Q = V_{\bar{i}}C_{max}$. At $C_t = C^* = 2C_{min}$, the transducer voltage reaches $2V_{\bar{i}}$ and the transducer network switches into a series configuration. Note that to make this switching possible, C_{max} must be greater than $2C_{min}$, which gives us a first condition for the existence of the steady-state operation. Starting from the moment at which $C_t = 2C_{min}$, the transducer gives its charges to the capacitive network C_1C_2. As C_t reduces from C^* to C_{min}, the charge given to C_1C_2 can easily be calculated if the series capacitive network C_1C_2 is considered as a single equivalent capacitor $C_{eq} = 1/(1/C_1 + 1/C_2)$ having a voltage $V_1 + V_2$. In this case, we can write for the charge $\Delta Q_{g\ i}$ given by the transducer at the ith cycle:

$$\Delta Q_{g\ i} = V_{\bar{i}}C_{max} - V_{\underline{i}}C_{min}, \qquad [10.26]$$

where $V_{t\underline{i}}$ is the transducer voltage (and the voltage on the capacitor C_{eq}) when the transducer capacitance is at its minimum. This voltage is calculated from the charge conservation law written for two moments: when $C_t = C_{min}$ (the left part of the equation) and when $C_t = C^*$ (the right part of the equation):

$$V_{t\underline{i}}(C_{min} + C_{eq}) = 2V_{\overline{i}}C_{eq} + V_{\overline{i}}C_{max} \qquad [10.27]$$

We obtain:

$$V_{t\underline{i}} = V_{\overline{i}}\frac{2C_{eq} + C_{max}}{C_{eq} + C_{min}} \qquad [10.28]$$

For $\Delta Q_{g\ i}$, we obtain:

$$\Delta Q_{g\ i} = V_{\overline{i}}C_{max} - V_{\overline{i}}\frac{2C_{eq} + C_{max}}{C_{eq} + C_{min}}C_{min} = V_{\overline{i}}C_{eq}\frac{C_{max} - 2C_{min}}{C_{eq} + C_{min}}. \qquad [10.29]$$

When $C_t = C_{min}$, the voltages of capacitors C_1 and C_2 are equal to:

$$\begin{aligned} V_{1\underline{i}} &= V_{\overline{i}} + \Delta Q_{g\ i}/C_1 \\ V_{2\underline{i}} &= V_{\overline{i}} + \Delta Q_{g\ i}/C_2 \end{aligned} \qquad [10.30]$$

C_t increases from C_{min} to C_{max}.

As C_t increases starting from C_{min}, the voltage on C_t decreases, and all diodes turn OFF. The transducer keeps its charge while decreasing its capacitance until its voltage drops to the value $\max(V_{1\underline{i}}, V_{2\underline{i}})$. According to our hypothesis $C_1 < C_2$, and hence, according to equation [10.30], $\max(V_{1\underline{i}}, V_{2\underline{i}}) = V_{1\underline{i}}$. The transducer voltage V_t reaches $V_{1\underline{i}}$ when $C_t = C^{**} = V_{t\underline{i}}C_{min}/V_{1\underline{i}}$. Starting from this moment, the diode D1 is ON, and C_1 is in parallel with C_t. Capacitor C_1 gives its charges to C_t, and the voltage of the both capacitors decreases. As this voltage drops to $V_{2\underline{i}}$, the diode D3 turns ON, the three capacitors are in parallel and their voltage continue to decrease till $C_t = C_{max}$. The switching on of the diode D3 happens for $C_t = C^{***}$, which can be calculated as:

$$(V_{t\underline{i}}C_{min} + V_{1\underline{i}}C_1)/(C^{***} + C_1) = V_{2\underline{i}} \qquad [10.31]$$

We leave it to the readers to prove that if $C_{max} \geq 2C_{min}$, the value C^{***} is below C_{max}. In other words, under the assumed hypotheses, the circuit we are analyzing enters into the mode where the voltages of the three capacitors are equal. The condition

$C_{max} > 2C_{min}$ was already mentioned necessary for the switching of the network into the series configuration; it is also a necessary condition for the three capacitors to become in parallel.

Starting from this point, the three capacitors are in parallel, they share their charges, and the calculation of the voltage at the beginning of the next cycle, when $C_t = C_{max}$ is straightforward. For this, it is enough to consider the charge balance:

– between $C_t = C_{max}$ and $C_t = C_{min}$, the transducer gives the charge $Q_{g\ i}$, and the capacitors C_1 and C_2 receive each the same charge $Q_{g\ i}$;

– as a result, between $C_t = C_{max}$ and the moment at which the three capacitors are in parallel, the variation of the sum of the charges of the capacitors is $-Q_{g\ i} + Q_{g\ i} + Q_{g\ i} = Q_{g\ i}$.

This point is paramount for understanding the operation of the Bennet's doubler: after a full cycle of operation, the sum of charges of the capacitors increases. This point is the main difference with the operation of the charge pump without flyback (Figure 10.3(b)), where the sum of charges of C_{res}, C_{st} and C_t is constant.

In this way, we have to add the charge $\Delta Q_{g\ i}$ to the charges that the capacitors had at the beginning of the cycle i, and to derive the voltage at the beginning of the ith cycle:

$$V_{\overline{i+1}}(C_1 + C_2 + C_{max}) = V_{\overline{i}}(C_1 + C_2 + C_{max}) + \Delta Q_{g\ i} \qquad [10.32]$$

Combining the last expression with [10.29], we obtain for the voltage of the three capacitors at the moment when $C_t = C_{max}$, i.e. at the beginning of the next $i + 1^{th}$ cycle:

$$\begin{aligned} V_{\overline{i+1}} &= V_{\overline{i}}\left[1 + \frac{C_{max} - 2C_{min}}{C_1 + C_2 + C_{max}} \cdot \frac{C_{eq}}{C_{eq} + C_{min}}\right] = \\ &= V_{\overline{i}}\left[1 + \frac{C_{max} - 2C_{min}}{C_1 + C_2 + C_{max}} \cdot \frac{C_1C_2}{C_1C_2 + C_{min}(C_1 + C_2)}\right] \end{aligned} \qquad [10.33]$$

We can now conclude our analysis: at the end of the ith cycle starting from the state given by [10.25], the voltages on the three capacitors became equal to $V_{\overline{i+1}}$. The last formula states that the next $i + 1^{th}$ cycle starts at a slightly higher voltage $V_{\overline{i+1}}$, and both the voltages and the charges of the capacitors increase exponentially with time. The increase is exponential since the coefficient near $V_{\overline{i}}$ in the recurrence equation [10.33] is above 1, as far as $C_{max} > 2C_{min}$.

This concludes the proof of the existence of a steady-state mode, according to the above definition. The inequality $C_{max} > 2C_{min}$ appears as a *sufficient* condition, since during the analysis we assumed it is valid. We let the readers to prove that this condition is also necessary.

The electrical energy gained by the network in one cycle in the steady-state mode is given by

$$\Delta W_i = \frac{1}{2}(C_1 + C_2 + C_{max})(V_{\overline{i+1}}^2 - V_{\bar{i}}^2). \qquad [10.34]$$

The gained energy is distributed over the three capacitors. If one of the capacitances C_1, C_2 or C_{max} dominate the others, *virtually all gained energy* is stored in this capacitor. This is very convenient for practical use of the Bennet's doubler: by defining $C_2 \gg C_1 \gg C_{max}$, the capacitor C_2 is a reservoir accumulating the converted energy, which can be easily interfaced with the load through an appropriate circuitry.

2) *QV cycle of the Bennet's doubler and approximated analysis in steady state*

Figure 10.12 presents the QV diagram corresponding to the operation of the transducer during an ith cycle. The QV diagram starts from the point D_i, which corresponds to $V_t = V_{\bar{i}}$, and the corresponding charge $Q_{t\bar{i}} = V_{\bar{i}} C_{max}$. The sections $D_iF_iB_i$ represent the evolution of the transducer as its capacitance decreases from C_{max} to C_{min}: D_iF_i corresponds to the constant charge operation, F_iB_i corresponds to the operation of the circuits when the capacitors are connected in series and the slope of the line is $-C_{eq}$.

The sections $B_iE_iE_i'D_{i+1}$ represent the evolution of the transducer as its capacitance increases from C_{min} to C_{max}. The section B_iE_i represents the evolution of the transducer with a constant charge given by $Q_{t\underline{i}}$. E_iE_i' corresponds to the evolution of the system when the capacitor C_1 is in parallel with C_t and the capacitor C_2 is disconnected by the diode D3, the slope of this line being $-C_1$. The section $E_i'D_{i+1}$ corresponds to the situation when the three capacitors are in parallel, and the slope is $-(C_1 + C_2)$. Note that the cycle is not closed: indeed, when the transducer capacitance is at C_{max} again, the charge and hence the voltage of all capacitors increased, and the starting point for the next cycle is at the point D_{i+1}, which is also situated on the line $C_t = C_{max}$, and its relative position is given by formula [10.33].

The obtained QV cycle is very close to an ideal rectangular cycle if the capacitors C_1 and C_2 are much larger comparing with C_{max}. Indeed, under this condition, the slopes of the lines F_iB_i, E_iE_i', $E_i'D_{i+1}$ are much larger than the slopes of the line $C_t = C_{max}$, and the cycle can be approximated by a rectangle. The area of the rectangle is given by the expression [10.34] in which C_{max} and C_{min} can be neglected in comparison to C_1, C_2 and C_{eq}. The simplified expression is given by:

$$\Delta W_i \approx V_{\bar{i}}^{\,2}(C_{max} - 2C_{min}). \qquad [10.35]$$

The energy ΔW_i is positive if $C_{min} - 2C_{max} > 0$. The readers may notice that the term $C_{max} - 2C_{min}$ also appeared when we defined the amount of charge $\Delta Q_{g\ i}$ that the transducer gives to the capacitor during the decrease of C_t. As well, $\Delta Q_{g\ i}$ is positive if $C_{max} > 2C_{min}$, since this was the condition allowing two fixed capacitors to become connected in series.

The assumption of C_1 and C_2 large compared to the transducer capacitance C_t allows a simple qualitative analysis of the circuit. If the capacitors are initially charged to the same voltages $V_1 = V_2 = V_0$, the exchange of charges between the fixed capacitors and the transducer does not modify substantially the voltages across C_1 and C_2, and at the time scale of a cycle, these voltages can be considered as being constant. In this case, in series configuration, the voltage applied to the transducer is $2V_0$, and in parallel V_0. Considering the general requirement applied to a transducer operating with a rectangular QV cycle discussed in section 10.1, such a mode is equivalent to $V_{res} = V_0$, $V_{st} = 2V_0$, and according to the condition [10.2], we conclude that C_{max} must be at least twice as large as C_{min}.

3) *Study of transient mode of the Bennet's doubler*

We showed that in the steady-state mode, the voltages on the three capacitors of the circuit are equal as C_t increases and approaches C_{max} (the segment $E_i'D_{i+1}$ of the QV cycle of Figure 10.12). In practice, the circuit may start from a state at which, for example, only C_1 is charged to a non-zero voltage. In this section, we analyze the evolution of the electrical state of the circuit when started from a state with any initial voltages on the capacitors, and we prove that the circuit always converges to the steady-state behavior described before.

For analysis of the transient process, we assume that at the ith cycle the voltages on the capacitors C_1 and C_2 are $V_{1\bar{i}} > V_{2\bar{i}}$, and without loss of generality of the demonstration, we suppose that at $C_t = C_{max}$ the transducer voltage has the voltage $\max(V_{1\bar{i}}, V_{2\bar{i}}) = V_{1\bar{i}}$. The readers may check that for any initial voltage of the transducer between $V_{1\bar{i}}$ and $V_{1\bar{i}} + V_{2\bar{i}}$ (the range allowed by the "voltage limiter" implemented by the capacitive network), the circuit reaches such a state in at most one cycle period. We now analyze the evolution of the capacitor voltages during one cycle.

C_t decreases from C_{max} to C_{min}.

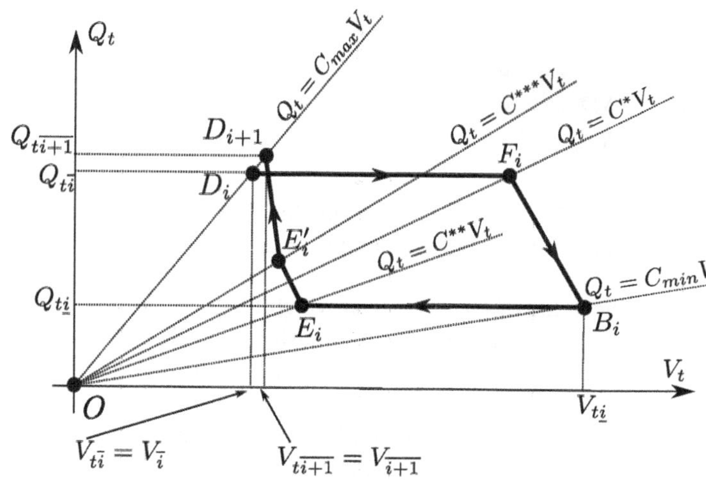

Figure 10.12. *Geometry of the QV diagram implemented by one cycle of the Bennet's doubler*

The evolution of the circuit is exactly as in the case where the initial voltages are equal, as analyzed in the previous section. The only difference is in the value C^* for which the diode D2 switches on and the capacitors C_1 and C_2 become connected in series. In the considered configuration,

$$C^* = C_{min}\frac{V_{1\bar{i}} + V_{2\bar{i}}}{V_{1\bar{i}}} \quad [10.36]$$

The transducer voltage as $C_t = C_{min}$ becomes:

$$V_{t\underline{i}}(C_{min} + C_{eq}) = (V_{1\bar{i}} + V_{2\bar{i}})C_{eq} + V_{1\bar{i}}C_{max} \quad [10.37]$$

and we obtain:

$$V_{t\underline{i}} = \frac{(V_{1\bar{i}} + V_{2\bar{i}})C_{eq} + V_{1\bar{i}}C_{max}}{C_{min} + C_{eq}} \quad [10.38]$$

The transducer gives to C_1 and C_2 the charge equal to:

$$\begin{aligned}\Delta Q_{g\,i} &= V_{t\bar{i}}C_{max} - V_{t\underline{i}}C_{min} = V_{1\bar{i}}C_{max} - V_{t\underline{i}}C_{min} \\ &= \frac{C_{eq}}{C_{min}+C_{eq}}\left[V_{1\bar{i}}(C_{max} - C_{min}) - V_{2\bar{i}}C_{min}\right].\end{aligned} \quad [10.39]$$

The evolution of V_1 and V_2 is given by:

$$V_{1\underline{i}} = V_{1\bar{i}} + \Delta Q_{g\ i}/C_1, \qquad [10.40]$$
$$V_{2\underline{i}} = V_{2\bar{i}} + \Delta Q_{g\ i}/C_2$$

C_t increases from C_{max} to C_{min}. During this phase, first C_1 then C_2 return charges to the transducer. Two scenarios are possible:

– *Scenario 1*: The value of $C_t = C^{***}$ at which C_2 is connected in parallel with C_1 and C_t is less than C_{max}. In this case, the considered cycle is the last cycle of the transient process, since the next cycle starts from $V_{1\overline{i+1}} = V_{2\overline{i+1}}$.

– *Scenario 2*: Otherwise, in the considered i^{th} cycle, C_2 never becomes in parallel with C_1. In this case, the capacitor C_2 keeps its charge received during the phase "from C_{max} to C_{min}", till $C_t = C_{max}$, and

$$V_{2\overline{i+1}} = V_{2\underline{i}}. \qquad [10.41]$$

Only C_1 gives its charge to C_t, starting from the moment they are in parallel (when $C_t = C^{**} < C_{max}$). The corresponding C^{**} value is calculated from the equation:

$$\frac{V_{t\underline{i}} C_{min}}{C_{**}} = V_{1\underline{i}}. \qquad [10.42]$$

and the final voltage of C_t and C_1 as $C_t = C_{max}$ is found from the conservation charge equation:

$$V_{t\underline{i}} C_{min} + V_{1\underline{i}} C_1 = V_{1\overline{i+1}}(C_{max} + C_1). \qquad [10.43]$$

The voltage $V_{t\overline{i+1}}$ is given by:

$$V_{t\overline{i+1}} = V_{1\overline{i+1}} = \frac{V_{t\underline{i}} C_{min} + V_{1\underline{i}} C_1}{C_{max} + C_1} = V_{1\bar{i}}. \qquad [10.44]$$

In the second scenario, the next cycle is also a transient process since $V_{1\overline{i+1}} \neq V_{2\overline{i+1}}$.

We conclude that during the transient process, from the cycle i to $i + 1$, the voltage V_2 increases whereas the voltage V_1 remains constant. The transient process ends when V_2 becomes equal to V_1.

The voltage $V_{\overline{2i+1}}$ can be expressed through $V_{1\bar{i}}$ and $V_{2\bar{i}}$ by combining equations [10.41], [10.40] and [10.39]:

$$V_{\overline{2i+1}} = V_{2\bar{i}}\left[1 - \frac{C_{eq}}{C_{min}+C_{eq}}\frac{C_{min}}{C_2}\right] + V_{1\bar{i}}\frac{C_{eq}(C_{max}-C_{min})}{(C_{min}+C_{eq})C_2} \qquad [10.45]$$

A simplified expression is obtained if we consider C_1 and C_2 much greater than C_{max}. In this case,

$$V_{\overline{2i+1}} \approx V_{2\bar{i}} + V_{1\bar{i}}\frac{C_{max}-C_{min}}{C_2} - V_{2\bar{i}}\frac{C_{min}}{C_2}. \qquad [10.46]$$

From this expression, it can be seen that the voltage V_2 increases, so approaching the voltage V_1. However, in order to enter into the steady-state mode corresponding to $V_1 = V_2$ at $C_t = C_{max}$, it can be checked that the inequality $C_{max} > 2C_{min}$ must be fulfilled. This can be proven by considering equation [10.46] as a linear recurrence relation with known and constant $V_{1\bar{i}}$, and by calculating the asymptotic value of V_2 as $i \to \infty$: if $C_{max} < 2C_{min}$, the asymptotic V_2 is less than V_1, and the steady-state mode as defined before cannot be reached.

The above analysis states that in the transient mode, the Bennet's doubler gains electrical energy, i.e. there is an electromechanical conversion. Indeed, comparing the voltages when $C_t = C_{max}$ at two neighboring cycles i and $i+1$, it can be seen that V_2 increases (see equation [10.46]), whereas the values of V_1 and V_t do not change. Moreover, as follows from equation [10.46], there is an energy gain even if $C_{max} < 2C_{min}$, as far as the values of V_1 and V_2 satisfy the basic condition for the existence of a rectangular QV cycle given by [10.2].

10.5.3. *Simulation of a Bennet's doubler*

In order to demonstrate the behavior of the Bennet's doubler in a realistic context, we implemented the model of the doubler as an Eldo netlist, which is given in Listing 10.1.

This netlist implements a Bennet's doubler with $C_1 = 10$ nF, $C_2 = 1$ nF, the transducer is a gap-closing transducer as presented in section 2.1 with $d_0 = 50\mu$m and $S = 10^{-4}$ m^2. The mobile electrode of the transducer moves according to sinusoidal law, with amplitude of $X_0 = 30$ μm and frequency $f = 100$ Hz. In this way, the transducer capacitance C_t varies in time according to the following law:

$$C_t(t) = \varepsilon_0 \frac{S}{d_0 - X_0 \sin(\omega t)}. \qquad [10.47]$$

From this equation, we calculate $C_{min} = 11.05$ pF and $C_{max} = 44.2$ pF. With these extreme values of C_t, the Bennet's doubler always reaches the steady-state mode, in which there is a sustainable energy conversion.

In this simulation, C_t varies according to the predefined law given by [10.47], so that the electromechanical coupling is not accounted for (see section 11.3).

```
***** The first line is ignored in spice netlists******

.model diode1 d level=1

**** The parameter definition *****
*The transducer gap
.param d=50e-6
*The transducer area
.param S=1e-4
*Permittivity of vacuum
.param eps0=8.85e-12
*Simulation time
.param simtime=30
*Amplitude of the mobile plane motion
.param X0=30e-6
*Frequency of the mobile plane motion
.param f=100

*The voltage source defining the trajectory of the mobile electrode
Vx x_value 0 sin 0 X0 f
*The voltage controlled voltage source
*defining the variation of the transducer capacitance
ECt Ct_value 0 value={eps0*S/(d-v(x_value))}

*The transducer
Ct b3 0 value={v(Ct_value)}

*The two fixed capacitances C1 and C2, initially charged
*to 10V and to 0V respectively
C1 b1 0 10e-9 ic=10
C2 b3 b2 1e-9 ic=0

*The three diodes
db1 0 b2 diode1
db2 b2 b1 diode1
db3 b1 b3 diode1

*Transient simulation command,
*with option uic (Use Initial Conditions)
```

```
.tran {simtime} {simtime} uic
*Plotting commands
.plot V(b1) V(b2) V(b3)
.plot V(b3,b2)
.plot V(b1,0)
.plot V(Ct_value)
*Writing of waveforms to text files , for the plot drawing
.printfile tran v(b1) file=V2.txt start=0 stop={simtime} step=5e-6
.printfile tran v(b3, b2) file=V1.txt start=0 stop={simtime} step=5e-6
.printfile tran v(Ct_value) file=Ct.txt start=0 stop={simtime} step=5e-6
.printfile tran v(b3) file=Vt.txt start=0 stop={simtime} step=5e-6

*simulator options
.options hmax=10e-6
```

Listing 10.1. *Bennet's doubler with variable capacitance: Eldo netlist*

Figure 10.1 presents a typical behavior of the Bennet's doubler with C_1 =10 nF, C_2 =1 nF, when initial voltages V_1 and V_2 are different, $V_2 > V_1$. The plot (a) presents a long-term evolution of the voltages on capacitors C_1 and C_2: during the transient, V_2 remains constant and V_1 increases. As the two voltages are equal, they evolve similarly, displaying an exponential increasing. The plot (b) presents a zoom of the time interval covering the transient and the beginning of the steady-state mode. The voltage on C_1 displays a very low ripple, comparing with the voltage on C_2. This is because of the difference between the value of the capacitances C_1 and C_2: the charge variation is the same on both (roughly $\Delta Q_{g\ i}$), but the voltage variation given by Q/C is greater on the smaller capacitor.

Figure 10.1(c) presents the voltages V_1, V_2 and the evolution of the transducer capacitance, so that the readers can observe the correlation between C_t variation and V_1, V_2 evolution during the transient process and in the steady state. At $C_t = C_{max}$, both voltages are equal in the steady-state mode. The plot (d) presents the evolution of the charge and voltage on the transducer in the same time interval as the plot (c). After the transient process, the max-to-min ratio of the voltage on the transducer is 2, and the voltage is constant when charge varies, and vice versa.

Figure 10.2(a) presents the QV cycles that the Bennet's doubler achieves in the steady state. The QV diagram is drawn for the cycle corresponding to the time instant 0.8 s. It can be noticed that the slope of the vertical sections of the QV cycle is non-infinite. However, the shape is very close to rectangular. This is because the slopes of the inclined segments are very large compared to C_{max}. In order to provide an insight into the operation of the circuit, the same network was simulated with smaller capacitors C_1 and C_2 (Figure 10.2(b)), whose values are now of the same order of magnitude as the transducer capacitance. The QV cycle obtained by simulation is now similar with the cycle of Figure 10.12 derived by the analysis.

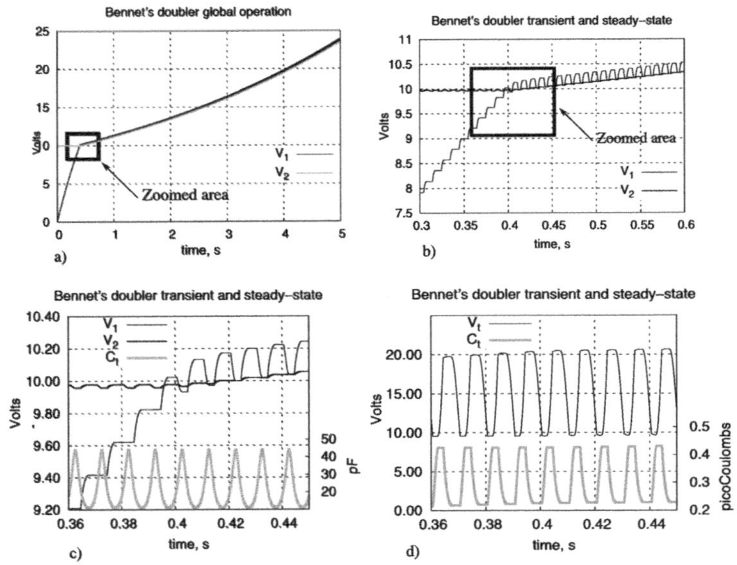

Figure 10.1. *Evolution of voltages in the Bennet's doubler: a) long-term evolution of the voltages V_1 and V_2, b) zoom on the time interval covering the end of the transient process and the beginning of the steady state, c) zoom of (b) presenting also the evolution of C_t, d) evolution of the transducer's voltage on the same time interval as in (c)*

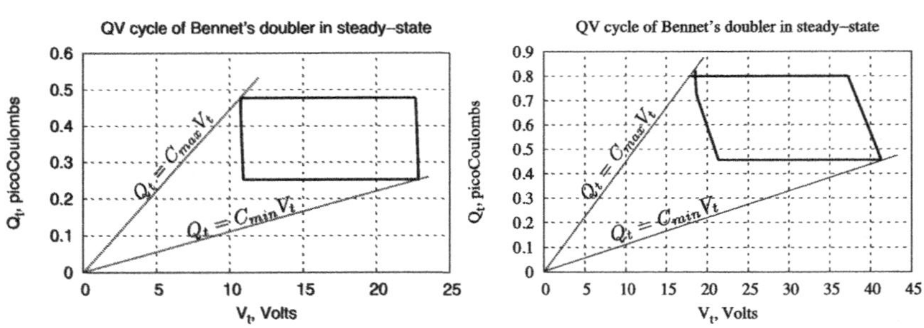

Figure 10.2. *QV diagram of Bennet's doubler obtained by simulation with the netlist given in listing 10.1, with different values of capacitors C_1 and C_2: a) $C_1 = 10nF$, $C_2 = 1nF$, as in the netlist given in listing 10.1, b) $C_1 = 500pF$, $C_2 = 100pF$. The corresponding C_{min} and C_{max} are 11.05 and 44.2 pF, respectively. It can be seen that when min$(C_1, C_2) \gg C_{max}$, the cycle is very close to be rectangular (a). As the fixed capacitances are of the same order of magnitude as the transducer capacitance, the cycle has the shape theoretically derived and shown in Figure 10.12*

11

Circuits Implementing Rectangular QV Cycles, Part II

In this chapter, we first present and analyze two circuits based on the diode rectifiers widely used with transducers biased by an electret, which were very briefly discussed in section 8.4.3. Then, we propose a discussion on a few practical aspects of the use of circuits implementing rectangular QV cycles: the effect of electromechanical coupling when these circuits are associated with a resonator, optimization of the power of energy conversion and the interfacing with load.

11.1. Analysis of the half-wave rectifier with a transducer biased by an electret

In this section, we analyze the conditioning circuit based on a half-wave diode rectifier depicted in Figure 8.9(a). We will demonstrate that the evolution of the electrical state of the transducer used with such a circuit follows a rectangular QV cycle. This allows us to analyze them from a standpoint of the theory developed in the previous chapter.

This network is electrically equivalent to the basic charge pump presented in Figure 10.3(a). This can be demonstrated if we equivalently transform the network of 8.9(a) by passing the voltage source V_{el} through the node 1 of the circuit (see Figures 11.1(a) and (b)) and then redraw the obtained topology (see Figure 11.1(c)). The similarity with the charge pump given in Figure 10.3(a) is obvious. The role of the voltage source V_{st} in Figure 10.3(a) is now played by the capacitor C_S put in series with the electret built-in voltage source V_{el}. The obtained charge pump pushes the charges from the voltage source V_{el} (on the left) toward the branch formed by the source V_{el} and the capacitor C_S (on the right). The charges are accumulated in C_S, so that the voltage V_S of this capacitor increases. The electrical state of the transducer evolves according to a rectangular QV cycle as shown in Figure 10.1, with the following parameters:

$$V_{res} = V_{el},\ V_{st} = V_{el} + V_S. \qquad [11.1]$$

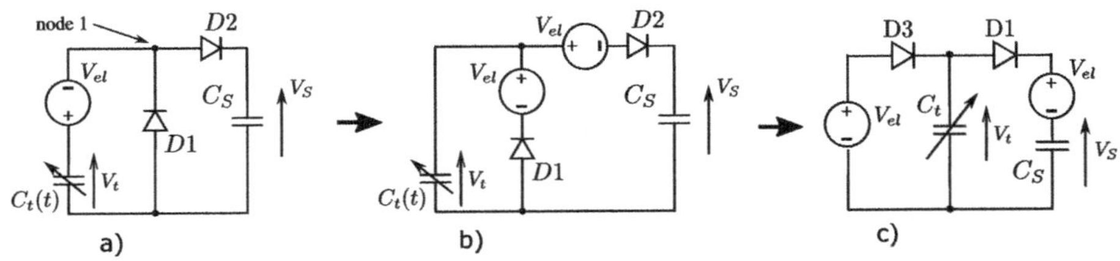

Figure 11.1. *Equivalent transform of the half-wave rectifier highlighting its analogy with a charge pump. The transformation steps are explained in the chapter body*

Using the theory of the rectangular QV cycles discussed in Chapter 10, we can calculate the key figures of performance of this network. For this calculation, we make the assumption of large value of C_S, such as $C_S \gg C_{max}$, which is usually valid in practical applications:

– the energy converted by one cycle is given by formula [10.1] merged with [11.1]:

$$\Delta W = V_{el} V_S C_{max} \left(1 - \frac{1 + V_S/V_{el}}{C_{max}/C_{min}}\right) \qquad [11.2]$$

– the last formula sets the upper bound on V_S. For ΔW to be non-negative, the following relation must hold:

$$V_S/V_{el} \leq C_{max}/C_{min} - 1, \qquad [11.3]$$

and the maximum allowed V_S called "saturation voltage" (see the discussion later in this section) is given by:

$$V_{S\ sat} = V_{el}(C_{max}/C_{min} - 1). \qquad [11.4]$$

– the optimal V_S value which yields a maximum of the converted energy, as far as V_{el}, C_{max} and C_{min} are fixed, is given by [10.3] merged with [11.1]:

$$V_{S\ opt} = \frac{1}{2} V_{el} \left(\frac{C_{max}}{C_{min}} - 1\right) \qquad [11.5]$$

and the corresponding ΔW_{max} is:

$$\Delta W_{max} = \frac{1}{4} V_{el}^2 C_{min} \left(\frac{C_{max}}{C_{min}} - 1 \right)^2 \qquad [11.6]$$

If $C_{res} \gg C_{max}$, the energy ΔW is mainly accumulated on the reservoir capacitor. Hence, after a cycle n, the output voltage V_S increases by a voltage ΔV_S, which can be calculated from the energy balance:

$$V_{S\ n+1}^2 C_{res}/2 = V_{S\ n}^2 C_{res}/2 + \Delta W_n, \qquad [11.7]$$

where $V_{S\ n}$ and ΔW_n are the output voltage and the converted energy at the nth cycle, and $\Delta V_n = \Delta V_{n+1} - \Delta V_n$. ΔW_n is calculated with equation [11.2] but substituting $V_{S\ n}$ instead of V_S. If the circuit starts up from the state at which $V_S = 0$, V_S increases after each cycle, till it reaches the saturation value $V_{S\ sat}$ defined in [11.4].

Figure 11.2(a) presents the QV cycles characterizing the evolution of the transducer electrical state, for different values of the output voltage V_S. It can be seen that, at extreme values of the output voltage 0 and $V_{S\ sat}$, the area of the cycles is close to zero, whereas there is an optimal output voltage yielding a QV cycle with a maximum area.

In the considered network of half-wave rectifier, the polarity of the electret built-in bias source V_{el} matters. We let the readers to prove that if the polarity of V_{el} is inverted in Figure 8.9(a), the saturation voltage of V_S is now:

$$V'_{S\ sat} = V_{el} \left(1 - \frac{C_{min}}{C_{max}} \right). \qquad [11.8]$$

As a result, the behavior of the modified circuit and the key figures of performances are different. In practice, this is not a problem, since the sign of the electret built-in voltage is defined by the sign of the charge embedded in the electret layer, and is known for a given practical device structure.

11.2. Analysis of the full-wave diode rectifier with transducer biased by an electret

Now, we propose to do the same analysis for the circuit in Figure 8.9(b). We will follow a similar methodology, transforming the network so as to emphasize its similarity with the basic charge pump of Figure 10.3(a) implementing a rectangular QV cycle.

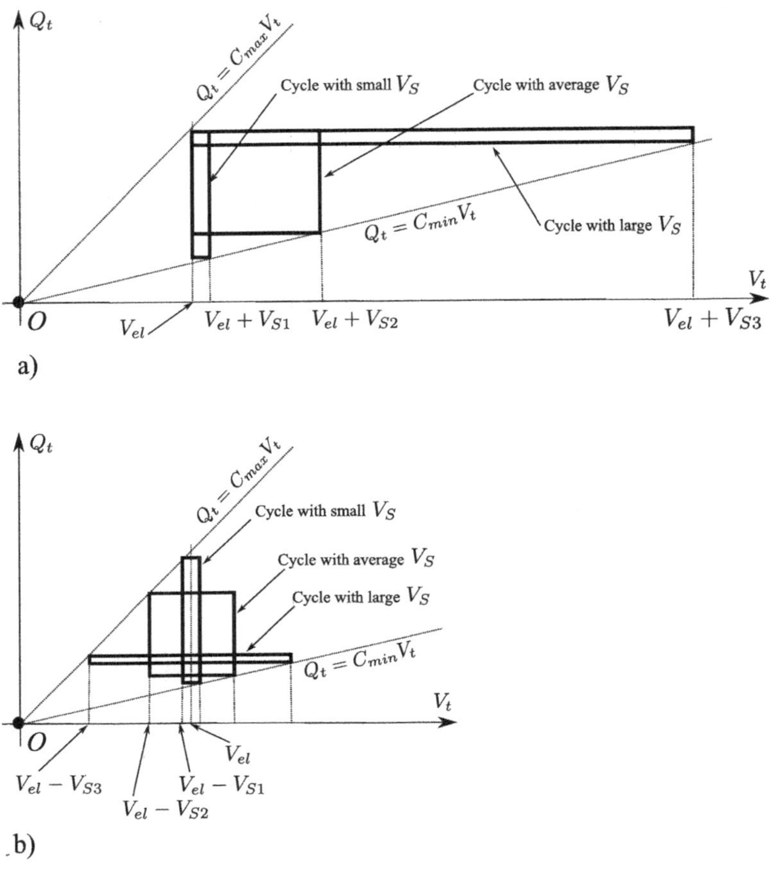

Figure 11.2. *QV cycles of the diode rectifiers used as conditioning circuits for transducer biased by the electret, for different values of the output voltage V_S: a) half-wave rectifier, b) full-wave rectifier. In both plots, V_{S3} is close to the saturation voltage*

Figure 11.3 presents a set of equivalent transformations of a simple conditioning circuit with a double wave diode rectifier. The first step (a $\rightarrow$ b) splits the voltage source V_{el} and the capacitors into two identical parts connected in series[1]. Note that the new components display exactly a half of the voltage across the original elements (e.g. the capacitors $2C_S$ both have the voltage $V_S/2$, etc.). The transform "b$\rightarrow$c" is possible due to the symmetry of the network: the nodes pointed by dotted arrows in

1 The readers should not be surprised that a half of the capacitor have twice the capacitance of the original capacitor: the two halves of the capacitor are connected in series.

Figure 11.3(b) have the same potentials, and as a result, they can be connected by a wire without affecting the electrical state of the network. The obtained circuit Figure 11.3(c) is then transformed by splitting each capacitor $2C_{res}$ into two parallel capacitors C_{res}. Each of them displays a voltage $V_S/2$. We obtain the circuit in Figure 11.3(d), which is composed of two identical networks (left and right), connected together in three corresponding nodes. Since the networks are identical, the voltage distribution is identical as well, and the connected nodes would have the same potential even without the connections. In other words, the wires which cross the border between the networks (the border being shown with a dotted line) have zero current.

The obtained half-circuit (Figure 11.3(e)) is very close to the charge pump given in Figure 10.3(a); an equivalent transformation of the voltage source gives the circuit in Figure 11.3(f), which is strictly equivalent to the charge pump in Figure 10.3(a). The corresponding V_{res} and V_{st} voltages are:

$$V_{res} = V_{el}/2 - V_S/2, \; V_{st} = V_{el}/2 + V_S/2, \qquad [11.9]$$

These voltages are the parameters of the QV cycle applied on the half of the transducer designated on the figure as $2C_t(t)$. Since it corresponds to half of the voltage across the transducer, the actual QV cycle will have the following parameters:

$$V_{res} = V_{el} - V_S, \; V_{st} = V_{el} + V_S, \qquad [11.10]$$

Now, it is straightforward to apply to this circuit the theory of the rectangular QV cycles, in order to obtain the key figures of performances:

– the energy converted by one cycle is given by formula [10.1] merged with [11.10]:

$$\Delta W = [(V_{el} - V_S)C_{max} - (V_{el} + V_S)C_{min}] \cdot 2V_S =$$

$$2V_S(C_{max} + C_{min}) \left[V_{el} \frac{C_{max} - C_{min}}{C_{max} + C_{min}} - V_S \right] \qquad [11.11]$$

– the last formula sets the upper bound on V_S. For ΔW to be non-negative, the following relation must hold:

$$V_S \leq V_{S\ sat}, \qquad [11.12]$$

where $V_{S\ sat}$ is the maximum V_S allowed for the energy conversion, which is the saturation voltage of the circuit. $V_{S\ sat}$ is given by:

$$V_{S\ sat} = V_{el}\frac{C_{max} - C_{min}}{C_{max} + C_{min}}, \qquad [11.13]$$

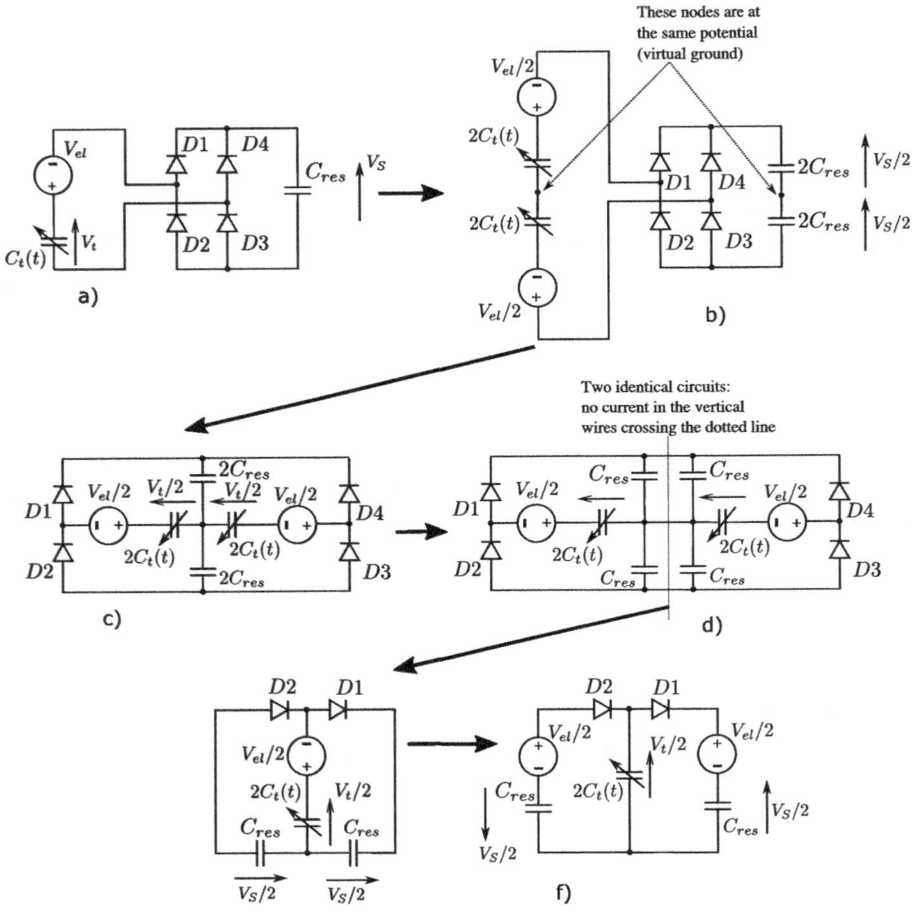

Figure 11.3. *Equivalent transform of the full-wave rectifier highlighting its analogy with a charge pump. The transformation steps are explained in the chapter body*

– the optimal V_S value which yields a maximum of the converted energy as far as V_{el}, C_{max} and C_{min} are fixed is given by [10.3] merged with [11.10]. This value is half of the saturation voltage:

$$V_{S\ opt} = \frac{1}{2}V_{S\ sat} = \frac{1}{2}V_{el}\frac{C_{max} - C_{min}}{C_{max} + C_{min}} \tag{11.14}$$

and the corresponding ΔW_{max} is:

$$\Delta W_{max} = \frac{1}{2}V_{el}^2 C_{min}\left(\frac{C_{max}}{C_{min}} - 1\right)^2 \cdot \frac{C_{min}}{C_{max} + C_{min}} \tag{11.15}$$

Figure 11.2(b) presents the QV cycles characterizing the evolution of the transducer electrical state, for different values of the output voltage V_S. As for the half-wave rectifier, the output voltage can be between 0 and $V_{S\ sat}$, and there is an optimal output voltage yielding a cycle with a maximum area.

At this point, the two considered circuits can be compared. In particular, the expressions [11.6] and [11.15] give the maximum energy convertible at one cycle, for the optimal value of the output voltage, with the given electret bias and the given C_{max} and C_{min}. It can be surprising that the full-wave rectifier highlights a lesser figure of maximum converted energy than the half-wave rectifier: the considered quantities are related with the factor $2C_{min}/(C_{min} + C_{max})$ always smaller than unity as far as $C_{min} < C_{max}$.

This disadvantage of the full-wave rectifier can be understood if we consider the expressions for the extreme values of the QV cycle of the two circuits, equations [11.1] and [11.9], as well as the QV diagrams of the two circuits. Both plots of Figure 11.2 are drawn with the same scale, with the same parameters of the circuit (electret voltage V_{el} and C_{max}, C_{min}), so as to allow a visual comparison. The width of the QV cycle of the full-wave rectifier is limited to $2V_{el}$ (since V_{res} cannot become negative), whereas the width of the QV cycle of the half-wave rectifier is defined by V_S limited by the saturation voltage, which, in turn, depends on the max-to-min ratio of the transducer capacitance. If C_{max}/C_{min} is large, we can see that the optimal value of V_S for the full-wave rectifier is of the same order of magnitude as V_{el} voltage, whereas for the half-wave rectifier the optimal V_S is 0.5 $V_{el}C_{max}/C_{min}$, i.e. the optimal V_S is "amplified" by the max-to-min ratio of the transducer capacitance. This is clearly observed on the QV diagrams.

However, the use of the full-wave rectifier may be advantageous in the case when the electret voltage is very high, and the technology does not allow us to set the voltage

V_S to the optimal value. Hence, we have $V_S \ll V_{el}$, and we have for the converted energy:

$$\Delta W_{max}^{\text{Half-wave}} \approx V_{el} V_S (C_{max} - C_{min}) \quad [11.16]$$

and

$$\Delta W_{max}^{\text{Full-wave}} \approx 2 V_S V_{el} (C_{max} - C_{min}) \quad [11.17]$$

Hence, under this particular condition of low V_S in comparison with the electret built-in voltage, the full-wave rectifier is twice as advantageous as the half-wave rectifier.

11.3. Dynamic behavior and electromechanical coupling of rectangular QV cycle conditioning circuits

The analysis of circuits considered in Chapters 8–10 was done in the electrical domain: it means that the variation of the capacitance was supposed to be independent of the electrical state of the transducer. As we mentioned in the previous chapters, in practice, this does not hold for real electrostatic transducers connected to microscale mechanical resonators. Indeed, as it was discussed in Chapter 6, the electromechanical coupling impacts the dynamic of the mobile mass vibrations. As a result, the extreme values of the transducer capacitance C_{max} and C_{min} can also vary with time.

All circuits considered in Chapters 10 and 11 operate in an autoaccumulative mode: the energy converted by the transducer is accumulated by the conditioning circuit. This accumulation leads to a modification of the dynamic biasing of the transducer, and hence, of the generated electrostatic force. For the calculation of the optimal circuit operation, the mechanical coupling has to be taken into account. Figure 11.4 presents a simulated evolution of V_{st} voltage of the charge pump in Figure 10.3, where the variation of the transducer capacitance is generated by the vibration of a linear mechanical resonator, whose parameters are given in the legend. The presence of the mechanical coupling can be attested by the evolution of the mobile mass amplitude *after the end of the mechanical transient*. As V_{st} voltage increases, the amplitude of the mass vibration changes slightly, and that has an impact of the variation on the transducer capacitance. The consequence of this phenomenon is a different evolution curve for $V_{st}(t)$ in comparison with the case where the amplitude of C_t variation was constant. Moreover, if the same circuit runs with different voltage V_{res}, the evolution of the amplitude of the mass vibration will be different.

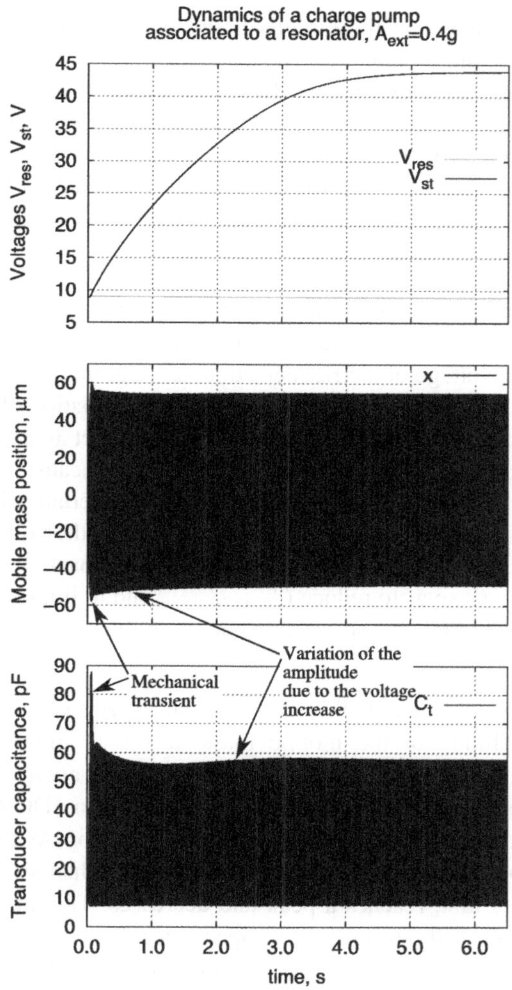

Figure 11.4. *Simulation of a charge pump (Figure 10.3(b)) where the variation of the transducer capacitance is done by a gap-closing transducer with* $S = 1 \cdot 10^{-4}$ *m*^{2}, $d_0 =$ 70 *µm, a resonator with* $m = 57 \cdot 10^{-6}$ *kg,* $k = 30.8$ *Nm*$^{-1}$ *and* $Q = 10$*, frequency of external vibrations is 120 Hz, acceleration amplitude is* 0.4 *g,* $C_{res} = 1$ *µF,* $C_{st} = 3$ *nF, initial voltage of* V_{res} *is* 9 *V*

A similar curve can be obtained for the circuits with electret layers, considered above in this chapter: the evolution of the voltage V_S impacts the QV cycle of the transducer, and modifies the amplitude of the mobile mass vibrations.

For the charge pump and the circuits based on rectifiers, the evolution of the output voltages with and without electromechanical coupling is qualitatively similar (compare the upper plots of Figures 10.5 and 11.4). What is impacted by the coupling is the optimal value of the output voltage at which the energy conversion is maximum: this value will be different from what is given by the formulas [10.3], [11.5] and [11.14]. Indeed, these formulas were derived under a hypothesis of fixed C_{max} and C_{min} values, whereas they vary with time in the case where the mechanical coupling is present.

However, a qualitative difference in behavior between a coupled and a purely electrical model is observed for the Bennet's doubler. Indeed, without mechanical coupling, the output voltage of the Bennet's doubler (e.g. the voltage V_1 or V_2, Figure 10.11) increases exponentially without any limit. If the variation of the capacitance is due to the motion of a mechanical system, we cannot expect an exponential evolution of the voltages and the energy converted at one cycle, because of the fundamental limit of the power which can be absorbed from the external vibrations, given by [3.37]. The mechanism practically limiting the increase in the voltage is the presence of the electrostatic force generated transducer, which impacts the amplitude of the mobile mass vibrations so as to reduce C_{max}/C_{min} and to limit the power converted by the transducer. Simulation and experiments highlight two possible behaviors of the system:

– *a "smooth" saturation*. In this case, as the energy of the Bennet's doubler increases, the amplitude of the mobile mass decreases smoothly, and the ratio C_{max}/C_{min} asymptotically approaches 2. This stops the energy conversion by the circuit. Such a behavior was experimentally observed in [DOR 14]. Figure 11.5 presents the simulation curves highlighting such a behavior. As the voltages V_1 and V_2 grow, the amplitude of the mobile mass vibration changes in a sophisticated manner: it increases first, reaches a peak and decreases. This mechanical dynamics impacts the variation of the transducer's capacitance. The evolution of the voltages is, in turn, impacted by the C_{max}/C_{min} ratio evolution: the rate is slow at the beginning, increases near the peak of the amplitude, decreases after the peak and becomes zero as C_{max}/C_{min} approaches 2 after time $t = 10$ s. The presence of a peak of amplitude at 2.4 s reminds a resonance phenomenon: indeed, we obtain a similar amplitude envelope if a linear resonator is excited by a sinusoidal signal with a frequency sweep. In our model, the frequency of the external acceleration is fixed, but the resonance frequency of the resonator is continuously modified by the voltages V_1 and V_2 responsible for the QV cycle definition (see discussion in section 6.2.5). The drift of the resonance frequency is clearly observed in the plot for the displacement x;

– *an "abrupt" saturation*. In this case, the increase in the output voltage does not produce the reduction of the amplitude the mobile mass, but creates conditions for the pull-in phenomena (see section 7.2.2.1). In this case, as the bias voltage of the circuit increases, the mobile mass sticks to the stoppers, and does not display any significant motion, so that C_{max}/C_{min} ratio is close to 1. This abruptly stops the increase in the output voltage. The resulting evolution is given in Figure 11.6.

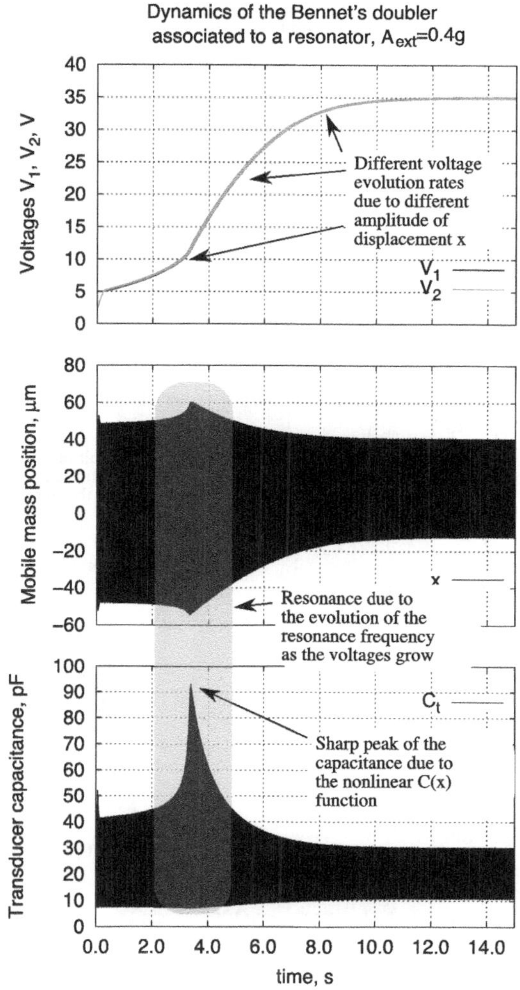

Figure 11.5. *Simulation results demonstrating a smooth saturation of a Bennet's doubler (Figure 10.11), where the variation of the transducer capacitance is done by a gap- closing transducer connected to a resonator, with the same parameters as for the plots of Figure 11.4. The frequency of the external vibrations is 110 Hz, acceleration amplitude is 0.4 g, C_1=10 nF, C_2=1 nF, initial voltage of V_1 is 5 V*

As a result, in practice, all considered circuits highlight a saturation of the output voltage, and for all circuits it is possible to find the value of the output voltage yielding a maximum energy conversion rate. A realistic design requires to take into account the mechanical coupling, which modifies the optimal values of the output voltages calculated in this chapter. This is discussed in the next section.

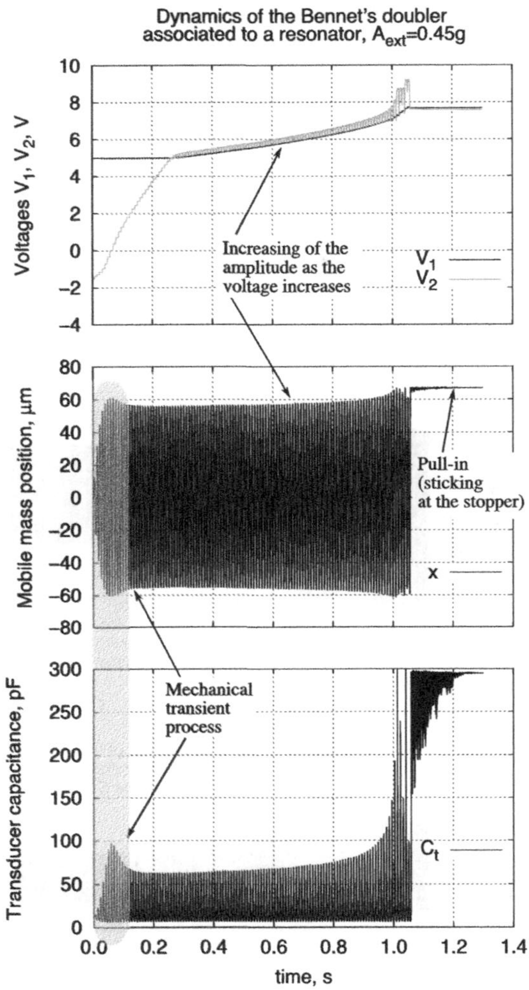

Figure 11.6. *Simulation results demonstrating an abrupt saturation of a Bennet's doubler (Figure 10.11), where the variation of the transducer capacitance is done by a gap-closing transducer connected to a resonator, with the same parameters as for the plots of Figure 11.4. The frequency of the external vibrations is 110 Hz, acceleration amplitude is 0.45 g, C_1 =10 nF, C_2=1 nF, initial voltage of V_1 is 5 V, the stopper is placed at 3 μm from the fixed electrode*

11.4. Practical use of conditioning circuits with rectangular QV cycle

The last question we would like to discuss is the following: how it is possible to maintain the output voltage of a conditioning circuit at the optimal level? The most straightforward and generic technique is presented in Figure 11.7.

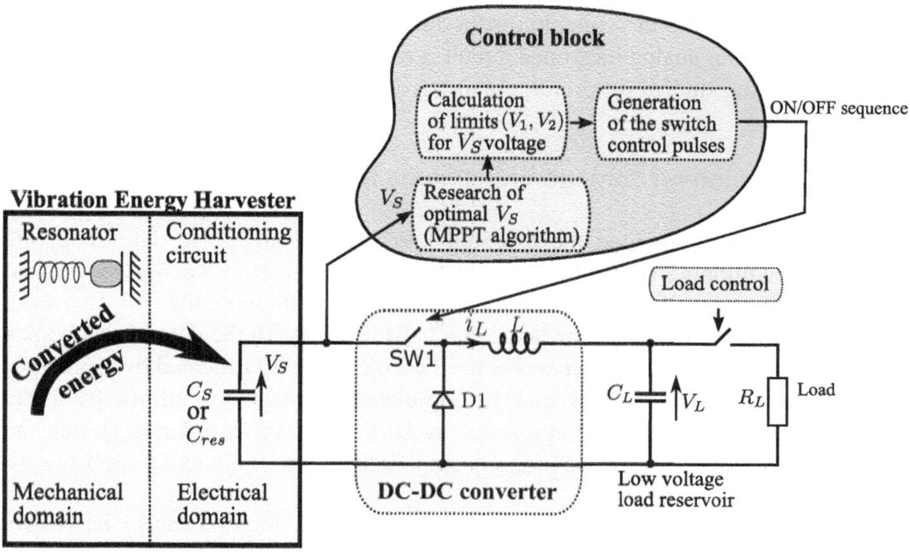

Figure 11.7. *Architecture implementing the control of the internal energy (the output voltage) of the vibration energy harvester in order to maximize the converted power*

This architecture is based on a DC-DC converter which interfaces the output capacitance of the conditioning circuit (C_S or C_{res} in the case of the charge pump with flyback) with a low-voltage energy reservoir available for the load supply. The output voltage of the harvester is defined by two energy flows: (1) the energy converted from the mechanical domain which tends to increase the output voltage, (2) the DC-DC conversion removing energy from the conditioning circuit to the load reservoir. By controlling the second flow, it is possible to control the average level of the output voltage, and to guarantee that it corresponds to the optimal value for the maximum power conversion. The technique is very similar with what we proposed for the control of the voltage V_{st} of the charge pump in section 10.4.2.2: the definition of the optimal voltage $V_{st\ opt}$, the definition of the interval (V_1, V_2) for the output voltage variation, the control of a DC-DC converter in order to remove the extra energy from the output capacitor, when necessary. The energy is accumulated on the load capacitor C_L and when there is enough energy for accomplishing some useful operation, the load is supplied. The most critical block is the one implementing the control of the DC-DC converter (represented as a gray cloud in the diagram), since a complex analog information processing is required, as discussed in section 10.4.2. A simplified version of such an interface with the load is proposed in [DE 11].

The implementation of this technique is a very challenging task, mainly because of the low level of power available for the implementation of the control algorithm. An ultra low power analog integrated circuit is required to make it possible. This issue is currently being studied in several research groups.

11.5. Conclusion on conditioning circuits for e-KEHs

Chapters 8–11 presented the essential knowledge available to date about the electrical conditioning of electrostatic transducers for vibration energy conversion. Such use of the capacitive transducer is relatively recent (no more than 10–15 years), and the associated practical knowledge is relatively small. To the date of this book's writing, most of the presented concepts have been essentially validated by prototypes issued from research projects and by simulation. Substantial efforts from the community of researchers and engineers are still necessary in order to elevate the maturity level of these techniques so as to employ them into commercial applications.

In many existing studies, the conditioning circuits of vibration energy harvesters have been analyzed without accounting for the mechanical coupling. However, both experiment and the theory emphasize that the electromechanical coupling deeply modifies the behavior of the circuit. In particular, whereas the electrical analysis claims that the bias voltage must be the largest possible in order to maximize the converted power (see, for instance, formulae [10.4] and [10.35]), analysis of the full system, simulations and experiment highlight the existence of an optimal value of the bias voltage [DOR 14]. Chapters 5 and 6 introduce analytical tools which can be used for the study of the coupling phenomena in capacitive energy harvesters, and provide some examples. The used mathematical tools are quite involved even for simple configurations, and to date, more investigation is required to have a comprehensive and handful representation of the effects of the electromechanical coupling.

Implementation of smart and adaptive behavior of capacitive KEHs like represented in Figure 11.7 and Figure 10.8(a) is mainly impeded by the low amount of power available from a microscale capacitive energy harvester. However, this limitation is specific to the technologies of integrated circuits (ICs) which have been available for the studies carried out to date. Future evolutions of the IC technologies will allow the implementation of the control circuitry with an acceptable power overhead. From this standpoint, electrostatic vibration energy harvesting may be seen as an emerging technology, having a strong potential for tomorrow's applications.

Bibliography

[ADA 98] ADAMS S.G., BERTSCH F.M., SHAW K.A. *et al.*, "Independent tuning of linear and nonlinear stiffness coefficients", *IEEE Journal of Microelectromechanical Systems*, vol. 7, pp. 172–180, 1998.

[AMR 11] AMRI M., BASSET P., COTTONE F. *et al.*, "Novel nonlinear spring design for wideband vibration energy harvesters", *Proceedings of PowerMEMS*, pp. 15–18, Seoul, South Korea, 2011.

[AND 87] ANDRONOV A.A., *Theory of Oscillators*, vol. 4, Courier Dover Publications, New York, 1987.

[AND 10] ANDÒ B., BAGLIO S., TRIGONA C. *et al.*, "Nonlinear mechanism in MEMS devices for energy harvesting applications", *Journal of Micromechanics and Microengineering*, vol. 20, no. 12, 2010.

[ANG 05] ANG S., OLIVA A., *Power-Switching Converters*, CRC Press, Boca Raton, FL, 2005.

[BAR 10] BARTSCH U., GASPAR J., PAUL O., "Low-frequency two-dimensional resonators for vibrational micro energy harvesting", *Journal of Micromechanics and Microengineering*, vol. 20, no. 3, 2010.

[BAR 12] BARTON D.A.W., BURROW S.G., CLARE L.R., "Energy harvesting from vibrations with a nonlinear oscillator", *Journal of Vibration and Acoustics*, vol. 132, pp. 91–97, 2012.

[BAS 09] BASSET P., GALAYKO D., PARACHA A.M. *et al.*, "A batch-fabricated and electret-free silicon electrostatic vibration energy harvester", *Journal of Micromechanics and Microengineering*, vol. 19, no. 11, November 2009.

[BAS 14] BASSET P., GALAYKO D., COTTONE F. *et al.*, "Electrostatic vibration energy harvester with combined effect of electrical nonlinearities and mechanical impact", *Journal of Micromechanics and Microengineering*, vol. 24, no. 3, 2014.

[BEE 09] BEEBY S.P., TUDOR M.J., WHITE N.M., "Energy harvesting vibration sources for microsystems applications", *Journal of Measurement Science and Technology*, vol. 17, pp. R175–R195, 2009.

[BEN 87] BENNET A., KAYE R., "An account of a doubler of electricity, or a machine by which the least conceivable quantity of positive or negative electricity may be continually doubled, till it becomes perceptible by common electrometers, or visible in sparks", *Philosophical Transactions of the Royal Society of London*, vol. 77, pp. 288–296, 1787.

[BLO 12] BLOKHINA E., GALAYKO D., HARTE P. *et al.*, "Limit on converted power in resonant electrostatic vibration energy harvesters", *Applied Physics Letters*, vol. 101, 2012.

[BLO 13] BLOKHINA E., GALAYKO D., BASSET P. *et al.*, "Steady-state oscillations in resonant electrostatic vibration energy harvesters", *IEEE Transactions on Circuits and Systems I*, vol. 60, pp. 875–884, 2013.

[BOI 10] BOISSEAU S., DESPESSE G., SYLVESTRE A., "Optimization of an electret-based energy harvester", *Smart Materials and Structures*, vol. 19, no. 7, 2010.

[BOI 13] BOISSEAU S., GASNIER P., GALLARDO M. *et al.*, "Self-starting power management circuits for piezoelectric and electret-based electrostatic mechanical energy harvesters", in *Journal of Physics: Conference Series*, vol. 476, 2013.

[BU 12] BU L., WU X., WANG X. *et al.*, "Liquid encapsulated electrostatic energy harvester for low-frequency vibrations", *Journal of Intelligent Material Systems and Structures*, vol. 24, no. 1, pp. 61–69, September 2012.

[COT 09] COTTONE F., GAMMAITONI L., VOCCA H., "Nonlinear energy harvesting", *Physical Review Letters*, vol. 102, 2009.

[COT 13] COTTONE F., BASSET P., GUILLEMET R. *et al.*, "Non-linear MEMS electrostatic kinetic energy harvester with a tunable multistable potential for stochastic vibrations", *Transducers & Eurosensors XXVII: The 17th International Conference on Solid-State Sensors, Actuators and Microsystems*, pp. 1336–1339, Barcelona, Spain, 2013.

[COT 14a] COTTONE F., BASSET P., MARTY F. *et al.*, "Electrostatic generator with free micro-ball and elastic stoppers for low-frequency vibration harvesting", *IEEE International Conference on Micro Electro Mechanical Systems (MEMS)*, pp. 385–388, San Francisco, CA, 26-30 January 2014.

[COT 14b] COTTONE F., FRIZZELL R., GOYAL S. *et al.*, "Enhanced vibrational energy harvester based on velocity amplification", *Journal of Intelligent Material Systems and Structures*, vol. 25, pp. 443–451, 2014.

[DAQ 10] DAQAQ M.F., "Response of uni-modal Duffing-type harvesters to random forced excitations", *Journal of Sound and Vibration*, vol. 329, pp. 3621–3631, 2010.

[DE 10] DE QUEIROZ A.C.M., "Electrostatic vibrational energy harvesting using a variation of Bennet's doubler", *53rd IEEE International Midwest Symposium on Circuits and Systems (MWSCAS)*, Seattle, USA, pp. 404–407, 2010.

[DE 11] DE QUEIROZ A.C.M., DOMINGUES M., "The doubler of electricity used as battery charger", *IEEE Transactions on Circuits and Systems II: Express Briefs*, vol. 58, no. 12, pp. 797–801, 2011.

[DEM 07] DE MARTINI B.E., RHOADS J.F., TURNER K.L. *et al.*, "Linear and nonlinear tuning of parametrically excited MEMS oscillators", *IEEE Jouranl of Microelectromechanical Systems*, vol. 16, pp. 310–318, 2007.

[DES 05a] DESPESSE G., Etude des phénomènes physiques utilisables pour alimenter en énergie électrique des micro-systèmes communicants, PhD Thesis, Institut National Polytechnique de Grenoble (INPG), 2005.

[DES 05b] DESPESSE G., JAGER T., JEAN-JACQUES C. *et al.*, "Fabrication and characterization of high damping electrostatic micro devices for vibration energy scavenging", *Proceedings of Design, Test, Integration and Packaging of MEMS and MOEMS (DTIP)*, pp. 386–390, Montreux, Switzerland, 2005.

[DET 11] DETERRE M., BOUTAUD B., DALMOLIN R. *et al.*, "Energy harvesting system for cardiac implant applications", *Symposium on Design, Test, Integration and Packaging of MEMS/MOEMS (DTIP)*, Aix-en-Provence, France, pp. 387–391, 2011.

[DOM 13] DOMINGUES M., DE QUEIROZ A.C.M., "Ultra-low-power control systems for electrostatic energy harvesters", *IEEE International Symposium on Circuits and Systems (ISCAS)* IEEE, Beijing, China, pp. 2960–2963, 2013.

[DOR 14] DORZHIEV V., KARAMI A., BASSET P. *et al.*, "Electret-free micromachined silicon electrostatic vibration energy harvester with the Bennet's doubler as conditioning circuit", *Electron Device Letters*, vol. 36, no. 2, pp. 183–135, 2014.

[DRA 13] DRAGUNOV V., DORZHIEV V., "Electrostatic vibration energy harvester with increased charging current", *Journal of Physics: Conference Series*, vol. 476, 2013.

[DUD 12] DUDKA A., GALAYKO D., BASSET P., "Design of controller IC for asynchronous conditioning circuit of an electrostatic vibration energy harvester", *IEEE International Conference on Internet of Things 2012 Workshop on Energy and Wireless Sensors*, Besancon, France, 2012.

[DUD 14] DUDKA A., GALAYKO D., BLOKHINA E. *et al.*, "Smart integrated conditioning electronics for electrostatic vibration energy harvesters", *IEEE International Symposium on Circuits and Systems (ISCAS)*, Melbourne, Australia, pp. 2600–2603, 2014.

[DUF 18] DUFFING G., *Erzwungene Schwingungen bei veränderlicher Eigenfrequenz und ihre technische Bedeutung*, Vieweg & Sohn, Braunschweig, Germany, 1918.

[ERI 07] ERICKSON R.W., MAKSIMOVIC D., *Fundamentals of Power Electronics*, Springer Science & Business Media, New York, 2007.

[ERT 09] ERTURK A., HOFFMANN J., INMAN D.J., "A piezomagnetoelastic structure for broadband vibration energy harvesting", *Applied Physics Letters*, vol. 94, 2009.

[FLO 11] FLORENTINO H.R., FREIRE R.C.S., SÁ A.V.S. *et al.*, "Electrostatic vibration energy harvester with piezoelectric start-up generator", *IEEE International Symposium on Circuits and Systems (ISCAS)*, Rio de Janeiro, Brazil, pp. 1343–1346, 2011.

[FU 14] FU B., BLOKHINA E., O'CONNELL D. *et al.*, "A wideband 2-DOF resonator for electromagnetic energy harvesting systems", *IEEE International Conference on Electronics, Circuits and Systems (ICECS)*, Marseille, France, pp. 878–881, 7–10 December 2014.

[GAL 05] GALAYKO D., KAISER A., LEGRAND B. *et al.*, "Tunable passband T-filter with electrostatically-driven polysilicon micromechanical resonators", *Sensors and Actuators A: Physical*, vol. 117, no. 1, pp. 115–120, 2005.

[GAL 07] GALAYKO D., PIZARRO R., BASSET P. *et al.*, "AMS modeling of controlled switch for design optimization of capacitive vibration energy harvester", *IEEE International Workshop on Behavioral Modeling and Simulation*, San Jose, USA, pp. 115–120, 2007.

[GAL 11a] GALAYKO D., BASSET P., "A general analytical tool for the design of vibration energy harvesters (VEHs) based on the mechanical impedance concept", *IEEE Transaction on Circuits and Systems I: Regular Papers*, vol. 58, no. 2, pp. 299–311, 2011.

[GAL 11b] GALCHEV T., KIM H., NAJAFI K., "Micro power generator for harvesting low-frequency and nonperiodic vibrations", *IEEE Journal of Microelectromechanical Systems*, vol. 20, no. 4, pp. 852–866, 2011.

[GAL 13a] GALAYKO D., BLOKHINA E., BASSET P. *et al.*, "Tools for analytical and numerical analysis of electrostatic vibration energy harvesters: application to a continuous mode conditioning circuit", *Journal of Physics: Conference Series*, vol. 476, 2013.

[GAL 13b] GALAYKO D., BLOKHINA E., "Nonlinear effects in electrostatic vibration energy harvesters: current progress and perspectives", *IEEE International Symposium on Circuits and Systems (ISCAS)*, Beijing, China, pp. 2719–2722, 2013.

[GAL 15] GALAYKO D., DUDKA A., KARAMI A. *et al.*, "Capacitive energy conversion with circuits implementing a rectangular charge-voltage cycle–part 1: analysis of the electrical domain", *IEEE Transactions on Circuits and System I*, vol. 62, no. 11, pp. 2652–2663, 2015.

[GRA 12] GRAFF K.F., *Wave Motion in Elastic Solids*, Courier Corporation, 2012.

[GUI 12] GUILLEMET R., BASSET P., GALAYKO D. *et al.*, "Design optimization of an out-of-plane gap-closing electrostatic vibration energy harvester (VEH) with a limitation on the output voltage", *Analog Integrated Circuits and Signal Processing*, vol. 71, no. 1, pp. 39–47, 2012.

[HAL 08] HALVORSEN E., "Energy harvesters driven by broadband random vibrations", *IEEE Journal of Microelectromechanical Systems*, vol. 17, pp. 1061–1071, 2008.

[HAL 13] HALVORSEN E., "Fundamental issues in nonlinear wideband-vibration energy harvesting", *Physical Review E*, vol. 87, 2013.

[HAR 14] HARTE P., BLOKHINA E., FEELY O. *et al.*, "Electrostatic vibration energy harvesters with linear and nonlinear resonators", *International Journal of Bifurcation and Chaos*, vol. 24, 2014.

[HIL 00] HILBORN R.C., *Chaos and Nonlinear Dynamics: An Introduction for Scientists and Engineers*, Oxford University Press, 2000.

[HIR 12] HIRSCH M.W., SMALE S., DEVANEY R.L., *Differential Equations, Dynamical Systems, and an Introduction to Chaos*, Elsevier Academic Press, New York, 2012.

[KEM 13] KEMPITIYA A., BORCA-TASCIUC D.-A., HELLA M.M., "Low-power interface IC for tri-plate electrostatic energy converters", *IEEE Transactions on Power Electronics*, vol. 28, no. 2, pp. 609–614, February 2013.

[KOR 00] KORN G.A., KORN T.M., *Mathematical Handbook for Scientists and Engineers: Definitions, Theorems, and Formulas for Reference and Review*, Courier Corporation, Dover Publications, New York, US, 2000.

[KRE 88] KREYSZIG E., *Advanced Engineering Mathematics*, John Wiley & Sons, New York, 1988.

[KUZ 02] KUZNETSOV A.P., KUZNETSOV S.P., RYSKIN N., *Nonlinear Oscillations*, Fizmatlit, Moscow, 2002.

[LAN 76] LANDAU L., LIFSHITZ E., *Mechanics: Volume 1 (Course Of Theoretical Physics)*, Elsivier Butterworth-Heinemann, Oxford, UK, 1976.

[LE 12] LE C.P., HALVORSEN E., "MEMS electrostatic energy harvesters with end-stop effects", *Journal of Micromechanics and Microengineering*, vol. 22, 2012.

[LEF 15] LEFEUVRE E., WEI J., MATHIAS H. *et al.*, "Single-switch inductorless power management circuit for electrostatic vibration energy harvesters", *Proceeding of IEEE NEWCAS Conference*, Grenoble, France, 2015.

[LI 06] LI H., PREIDIKMAN S., BALACHANDRAN B., MOTE JR C., "Nonlinear free and forced oscillations of piezoelectric microresonators", *Journal of Micromechanics and Microengineering*, vol. 16, p. 356, 2006.

[LIM 06] LIM L., BASSET P., PARACHA A.M. *et al.*, "Design of a novel highly capacitive MEMS-based vibration to electricity energy converter with suspended in-plane overlap technology", *Proceedings of Asia-Pacific Conference of Transducers and Micro-Nano Technology*, Singapore, 2006.

[LU 15] LU Y., COTTONE F., MARTY F. *et al.*, "Low-frequency MEMS electrostatic vibration energy harvester with corona-charged vertical electret and nonlinear stoppers", *PowerMEMS*, Boston, USA, 2015.

[MEN 01] MENINGER S., MUR-MIRANDA J., AMIRTHARAJAH R. *et al.*, "Vibration-to-electric energy conversion", *IEEE Transactions on Very Large Scale Integration (VLSI) Systems*, vol. 9, pp. 64–76, 2001.

[MIK 10] MIKI D., HONZUMI M., SUZUKI Y. *et al.*, "Large-amplitude MEMS electret generator with nonlinear spring", *Proceedings of IEEE Conference on Microelectromechanical Systems (MEMS)*, Wanchai, Hong Kong, pp. 176–179, 24–28 January 2010.

[MIT 08] MITCHESON P.D., STERKEN T., HE C. *et al.*, "Electrostatic microgenerators", *Measurement and Control*, vol. 41, no. 4, pp. 114–119, 2008.

[MIT 12] MITCHESON P.D., GREEN T.C., "Maximum effectiveness of electrostatic energy harvesters when coupled to interface circuits", *IEEE Transaction on Circuits and Systems I: Regular Papers*, vol. 59, no. 12, pp. 3098–3111, 2012.

[MUR 04] MUR-MIRANDA J.O., Electrostatic vibration-to-electric energy conversion, PhD Thesis, MIT, 2004.

[NAR 09] NARUSE Y., MATSUBARA N., MABUCHI K. *et al.*, "Electrostatic micro power generation from low-frequency vibration such as human motion", *Journal of Micromechanics and Microengineering*, vol. 19, no. 9, pp. 19–22, 2009.

[NAY 93] NAYFEH A., *Introduction to Perturbation Techniques*, Wiley, New York, 1993.

[NAY 05] NAYFEH A.H., YOUNIS M.I., ABDEL-RAHMAN E.M., "Reduced-order models for MEMS applications", *Nonlinear Dynamics*, vol. 41, nos. 1–3, pp. 211–236, 2005.

[NAY 08a] NAYFEH A.H., BALACHANDRAN B., *Applied Nonlinear Dynamics*, vol. 24, Wiley-VCH, New York, 2008.

[NAY 08b] NAYFEH A.H., MOOK D.T., *Nonlinear Oscillations*, John Wiley & Sons, New York, 2008.

[NEJ 92] NEJMARK I., LANDA P., *Stochastic and Chaotic Oscillations*, vol. 77, Springer, 1992.

[NGU 10] NGUYEN S.D., HALVORSEN E., JENSEN G. *et al.*, "Fabrication and characterization of a wideband MEMS energy harvester utilizing nonlinear springs", *Journal of Micromechanics and Microengineering*, vol. 20, 2010.

[NGU 11] NGUYEN S.D., HALVORSEN E., "Nonlinear springs for bandwidth-tolerant vibration energy harvesting", *IEEE Journal of Microelectromechanical Systems*, vol. 20, pp. 1225–1227, 2011.

[O'RI 14] O'RIORDAN E., BLOKHINA E., FEELY O. *et al.*, "Modelling and analysis of vibration energy harvesters with charge pump conditioning circuits", *IEEE International Symposium on Circuits and Systems (ISCAS)*, pp. 1247–1250, Melbourne, Australia, 1–5 June 2014.

[O'RI 15] O'RIORDAN E., DUDKA A., GALAYKO D., *et al.*, "Capacitive energy conversion with circuits implementing a rectangular charge-voltage cycle. Part 2: electromechanical and nonlinear analysis", *IEEE Transaction on Circuits and Systems I*, vol. 62, no. 11, pp. 2664–2673, 2015.

[PAR 07] PARACHA A.M., BASSET P., MARTY F. *et al.*, "A high power density electrostatic vibration-to-electric energy converter based on an in-plane overlap plate (IPOP) mechanism", *Symposium on Design Test Integration and Packaging of MEMS/MOEMS (DTIP)*, Lago Maggiore, Italy, 2007.

[PAR 09] PARACHA A.M., BASSET P., GALAYKO D. *et al.*, "MEMS DC/DC converter for 1D and 2D vibration-to-electricity power conversion", *International Solid-State Sensors, Actuators and Microsystems Conference, TRANSDUCERS 2009*, Denver, CO, pp. 2098–2101, 2009.

[PAR 14] PARASURAMAN J., SUMMANWAR A., MARTY F. *et al.*, "Deep reactive ion etching of sub-micrometer trenches with ultra high aspect ratio", *Microelectronic Engineering*, vol. 113, pp. 35–39, 2014.

[PET 82] PETERSEN K.E., "Silicon as a mechanical material", *Proceedings of the IEEE*, vol. 70, no. 5, pp. 420–457, 1982.

[RAB 89] RABINOVICH M.I., *Oscillations and Waves in Linear and Nonlinear Systems*, Taylor & Francis, 1989.

[REN 13] RENAUD M., ALTENA G., GOEDBLOED M. *et al.*, "A high performance electrostatic MEMS vibration energy harvester with corrugated inorganic SiO 2-Si 3 N 4 electret", *Transducers & Eurosensors XXVII: The 17th International Conference on Solid-State Sensors, Actuators and Microsystems* Barcelona, Spain, pp. 693–696, 2013.

[RIL 06] RILEY K., HOBSON P., BENCE S., *Mathematical Methods for Physics and Engineering: A Comprehensive Guide*, Cambridge University Press, Cambridge, UK, 2006.

[RIS 14] RISQUEZ S., WEI J., WOYTASIK M. *et al.*, "Self-biased inductor-less interface circuit for electret-free electrostatic energy harvesters", *Journal of Physics: Conference Series*, vol. 557, pp. 12052–12056, 2014.

[ROU 02] ROUNDY S., WRIGHT P., PISTER K., "Micro-electrostatic vibration-to-electricity converters", *Proceedings ASME International Mechanical Engineering Congress*, New Orleans, LA, 17–22 November 2002.

[SEN 01] SENTURIA S.D., *Microsystem Design*, Kluwer Academic Publishers, Boston, MA, 2001.

[SMI 54] SMITH C.S., "Piezoresistance effect in germanium and silicon", *Physical Review*, vol. 94, no. 1, pp. 42–49, 1954.

[STA 10] STANTON S.C., MCGEHEE C.C., MANN B.P., "Nonlinear dynamics for broadband energy harvesting: Investigation of a bistable piezoelectric inertial generator", *Physics D: Nonlinear Phenomena*, vol. 239, pp. 640–653, 2010.

[STE 03] STERKEN T., FIORINI P., BAERT K. *et al.*, "An electret-based electrostatic/ microgenerator", *12th International Conference on Transducers, Solid-State Sensors, Actuators and Microsystems*, Boston, USA, vol. 2, pp. 1291–1294, 2003.

[SUZ 11] SUZUKI Y., "Recent progress in MEMS electret generator for energy harvesting", *IEEJ Transactions on Electrical and Electronic Engineering*, vol. 6, no. 2, pp. 101–111, 2011.

[TAK 15] TAKAHASHI T., SUZUKI M., NISHIDA T. *et al.*, "Vertical capacitive energy harvester positively using contact between proof mass and electret plate–Stiffness matching by spring support of plate and stiction prevention by stopper mechanism", *IEEE International Conference on Microelectromechanical systems*, Estoril, Portugal, pp. 1145–1148, 18–22 January 2015.

[TAN 10] TANG L., YANG Y., SOH C.K., "Toward broadband vibration-based energy harvesting", *Journal of Intelligent Material Systems and Structures*, vol. 21, pp. 1867–1897, 2010.

[TOR 09] TORRES E.O., RINCON-MORA G.A., "Electrostatic energy-harvesting and battery-charging CMOS system prototype", *IEEE Transaction on Circuits and Systems I: Regular Papers*, vol. 56, pp. 1938–1948, 2009.

[TRU 01] TRUBETSKOV D.I., ROZHNEV A.G., *Linear Oscillations and Waves*, Fizmatlit, Moscow, 2001.

[TVE 10] TVEDT L.G.W., NGUYEN D.S., HALVORSEN E., "Nonlinear behavior of an electrostatic energy harvester under wide-and narrowband excitation", *IEEE Journal of Microelectromechanical Systems*, vol. 19, pp. 305–316, 2010.

[VOC 14] VOCCA H., COTTONE F., "Kinetic energy harvesting", in FAGAS G. (ed.), *ICT – Energy – Concepts Towards Zero – Power Information and Communication Technology*, available at: http://www.intechopen.com/books/ict-energy-concepts-towards-zero-power-information-and-communication-technology/kinetic-energy-harvesting, 2014.

[YEN 06] YEN B.C., LANG J.H., "A variable-capacitance vibration-to-electric energy harvester", *IEEE Transaction on Circuits and Systems–I: Regular Papers*, vol. 53, no. 2, pp. 288–295, 2006.

[ZHU 10] ZHU D., TUDOR M.J., BEEBY S.P., "Strategies for increasing the operating frequency range of vibration energy harvesters: a review", *Journal of Measurement Science and Technology*, vol. 21, 2010.

Index

A, B, C

D, E

F, G, H

I, L, M, N

O, P, R, S

T, U, V

Other titles from

iSTE

in

Nanoscience and Nanotechnology

2015

ROSSI Carole
Al-based Energetic Nanomaterials

LEPRINCE-WANG Yamin
Piezoelectric ZnO Nanostructure for Energy Harvesting

2014

BALESTRA Francis
Beyond CMOS Nanodevices 1
Beyond CMOS Nanodevices 2

LACAZE Pierre Camille, LACROIX Jean-Christophe
Non-volatile Memories

LATU-ROMAIN Laurence, OLLIVIER Maelig
Silicon Carbide One-dimensional Nanostructures

NICU Liviu, LEÏCHLE Thierry
Micro- and Nanoelectromechanical Biosensors

SENEZ Vincent, THOMY Vincent, DUFOUR Renaud
Nanotechnologies for Synthetic Super Non-wetting Surfaces

TRIPON-CANSELIET Charlotte, CHAZELAS Jean
Nanoscale Microwave Engineering: Optical Control of Nanodevices

2012

BELLEVILLE Marc, CONDEMINE Cyril
Energy Autonomous Micro and Nano Systems

CLAVERIE Alain
Transmission Electron Microscopy in Micro-nanoelectronics

LACAZE PIERRE Camille
Nanotechnologies: Concepts, Processing and Applications

2010

BALESTRA Francis
Nanoscale CMOS / Innovative Materials, Modeling and Characterization

QUERLIOZ Damien, DOLLFUS Philippe
The Wigner Monte Carlo Method for Nanoelectronic Devices: Particle Description of Quantum Transport and Decoherence

TAKADOUM Jamal
Nanomaterials and Surface Engineering

VODA Alina
Micro, Nanosystems and Systems on Chips: Modeling, Control and Estimation

2008

DAVIM J. Paulo, JACKSON J.
Nano and Micromachining

NOUAILLHAT Alain
An Introduction to Nanoscience and Nanotechnology

OUISSE Thierry
Electron Transport in Nanostructures and Mesoscopic Devices

2006

RIGNEAULT Hervé *et al.*
Nanophotonics

Printed and bound by CPI Group (UK) Ltd, Croydon, CR0 4YY

13/09/2026

14961079-0002